EVOLVING THE MODERN BREASTFEEDING EXPERIENCE: HOLISTIC LACTATION CARE IN THE FIRST 100 HOURS

Christine Staricka

ISBN-979-8-9904583-1-4

Cover design by: Christine Staricka
Library of Congress Control Number: 2018675309
Printed in the United States of America

To Katie, Jillian, and Alli, you are everything.
Thank you for being on my team - we're changing the world!

CONTENTS

PROLOGUE

It is my intention that reading this book will allow you to make a shift - or a series of shifts - in how you think about breastfeeding and lactation.

That shift will allow you to provide more effective clinical lactation care and breastfeeding support.

Here's why: several monumental changes have occurred over the span of my own involvement in the realm of lactation support, both as a mother and as a lactation care provider.

They have fundamentally changed how we do this work, what is required of us, and our potential to impact others.

Among them: access to the internet, the proliferation of products, devices, and supplements marketed to support lactation, legislative change in the United States that led to every birth event making a mother eligible to receive a breast pump, and a global pandemic that confined people to their homes and reduced access to many types of social support and health and wellness care.

Of course, many other incredibly important changes have occurred that are shaping our field and its future.

But it is those major ones that most clearly define for me what is different today than when I first went from a breastfeeding mother regularly attending a breastfeeding support group to a student in a lactation educator course.

The mindset shift in The First 100 Hours will also help you better understand and frame your own experiences around lactation.

As you read - and this is a different kind of lactation book than you are probably used to - I will challenge you and ask you to engage with your academic, clinical, personal, emotional, and theoretical thinking systems.

One element of change in the field of human lactation care that may be challenging to some is the concept that some people who give birth and breastfeed do not identify themselves as mothers.

I want to tell you about a patient I once worked with in the hospital.

I worked with her because I was the only person who could.

You see, she had insulted, offended, and kicked out of her room every other hospital staff member who tried to talk to her that day, including the registered nurse who was legally responsible for documenting her care.

She created a scene, and she made my coworkers cry.

I didn't want to go in there.

I'm very sensitive to confrontation.

I wanted no part of that.

I didn't want her to think I was on her side, and I really didn't want to have to talk to her support person, whose opinions were tattooed on his body in the form of a swastika.

She only let me come in because I am white.

She wasn't the first or last client I have provided lactation care to whose feelings, opinions, behaviors, preferences, or general needs made me uncomfortable.

Ethically, it was clear to me that I was responsible for ensuring that this patient received lactation care, and as a hospital employee, it was also my job, especially because she had mistreated my coworkers and prevented them from doing their jobs.

I was able to talk to her about breastfeeding so we could at least document that in the baby's chart, and I tried to build a bridge so her nurse could come in and give her the care she was legally responsible for providing.

I'm not saying I saved the day.

I'm telling you this because I want you to be clear that there will be people who cross your path during their pregnancies and early parenting journeys that do not fit what you hope all mothers and fathers will be, whatever that looks like.

Some people whose babies feed from their breasts do not call it breastfeeding.

Some people who have breasts that lactate do not refer to them as breasts.

Some people will tell you about cultural traditions around pregnancy, babies, and postpartum that you have never heard of or don't believe are necessary or logical.

What an awesome opportunity to learn about other people!

Throughout this book, I will mostly refer to pregnant and lactating folks as "mothers" for consistency and clarity because I don't think that everyone brings the same level of understanding around this.

In one-to-one situations, I use the nicknames, pronouns, and terms people ask me to use. For example, if they ask me to remove my shoes before entering for a home visit because it is their family's practice, I will do that as well.

If we all accept that the work we are doing in helping people who are pregnant and breastfeeding is already fraught with enough structural barriers, we must then see to it that we don't create more through our actions or behaviors.

Most lactation care providers I meet have committed themselves to improving the environment and landscape in which mothers pursue their lactation goals, and they are primarily effecting that change through education, clinical work, and advocacy.

I spent the first part of my lactation career doing exactly those things. Now, I have committed myself to ensuring that we as lactation care providers, as well as others who are asked to provide support in conjunction with the other types of care and education they perform through other disciplines, have an abundance of tools and resources to scaffold our didactic knowledge acquisition.

I have developed methods to continue our growth as counselors and educators.

I do this work because I have seen how important it is to protect our emotional well-being so that we can be healthy and resilient and continue our work for as long as we choose.

I have had the privilege of speaking about lactation and the lactation career with so many people from many different settings:

folks who came through all the various pathways from many types of careers and backgrounds

people in different geographic regions around the world

people with economic resources to support their dreams and people who lacked them

people whose skin color, religion, number of children, relationship status, or gender identity made entry into the field more challenging

people who came in when computers weren't in every pocket and people who are digital natives who have been on the internet since they were children

people who breastfed their children and people who couldn't

and so many more.

I love talking to people about this work.

I have always made it a point to take every opportunity to ask what brought people here, to hear about their passion or fascination with human lactation, and to listen to them talk about how they think it could be if we could just change X and Y.

I have had many of these opportunities because I have worked and volunteered in organizations serving everything from local providers to national and international associations.

When I go to lactation conferences, I spend my time outside the presentations I give talking to people.

Whenever I have the chance to give a presentation, the best part is right afterward when people ask questions and give feedback.

I learn so much from getting people to talk about why and how they do this work, and I want to make it better for them—and for you, reader.

I want you to know how much it matters to the world that YOU do it.

This book should help you see that there is not just a skeleton holding the lactation workforce together, but fascia, intertwined in ways that weave YOU in.

You are a valuable piece of the whole.

I want you to recognize when you could really use the support of your peers to get through something, and I want you to keep your eyes open for when you can be that support for your peers.

This book is for you, to show you what we can do together with some shifts in thinking about lactation and the lactation career path.

I have seen the many ways that lactation care providers can struggle with:

Moving from another primary healthcare discipline, especially nursing, into lactation care as their focus

their transition to clinical work from an education-centered book.

Let's get started!

INTRODUCTION

T he modern American breastfeeding experience often looks like this: limited prenatal education about breastfeeding and lactation, labor and birth with a high number of interventions, a rushed and overwhelming postpartum stay in the hospital with limited lactation support, and a whirlwind discharge back home into the community with a thin amount of information about how to get help with anything related to their health, let alone breastfeeding.

It is left up to the brand-new mother and/or their partner to figure out what help they might need, how to get it, and, often, how to pay for it.

New parents have been flooded with messages from people in real life and online throughout the pregnancy and now that their baby has been born.

These messages convey a strong theme: having a baby means you need to acquire many things.

Many of these "things" are related to feeding the baby, and many are relatively costly.

Aside from the products, they are also warned about how hard feeding can be, especially breastfeeding, which they are warned will be heavily pushed by everyone they meet.

Companies positioning themselves to profit from the new family highly target expectant and new parents, as well as healthcare providers, through marketing efforts.

Healthcare providers profit as well by receiving gifts in exchange for their endorsement (obvious or implicit) of the products, and they are encouraged to do so while feeling good about how they are removing "pressure" to breastfeed from their patients or clients.

Mothers often tell us that their OB/GYN did not say much about breastfeeding at all and that when they ran into problems or doubts about breastfeeding, the nurses in the hospital and their baby's pediatrician encouraged them to "just pump" or "just give some formula" without offering much support or help with breastfeeding.

Many new mothers say that the people in their lives also nudged them away from breastfeeding under the guise of helping make everything easier.

A popular narrative currently interfering with mothers being able to meet their own breastfeeding goals is the inaccurate messaging being propagated about the interplay between breastfeeding and a mother's mental health.

Add to this that the new mother has been minimally supported to do everything else required in their lives, including physical recovery.

The lack of paid family leave means that women are often the sole caregivers for other children and/or disabled or older members of the household while their partners are out of the house at work.

Then they, themselves, return to work far too early to accommodate their physical recovery from childbirth.

Running the household generally falls to them, and they are out driving around or riding public transportation to do errands, get themselves and their baby to doctor appointments, and more.

The postpartum experience is less the "cuddle and cocoon" time that all new parents need and more the "figure out how to solve all these problems" time that increases their stress levels, making sleep deprivation feel much worse and making infant feeding fraught and challenging.

After all, it's basically impossible to follow your baby's cues and meet their needs for feeding and sleeping if you're unable to sit and focus on the baby because you are trying to meet everyone else's needs.

No individual parent or family is to blame for this modern

American breastfeeding experience.

The choices that people make about infant feeding in our society are a reflection of the environment.

Overcoming the messaging and influence of bad information is incredibly difficult.

When people don't know that the information surrounding them is bad, they can't be blamed for what they do with that information.

But the biological reality remains the same.

Breastfeeding and human milk are still the most important immunological defense and nutritional foundation available to a newborn.

When I created the First 100 Hours concept and program in 2014, my goal was to develop an entire system of educating parents and healthcare workers with information and understanding of the fundamentals of human lactation that could transcend the cultural myths and unhelpful practices that plague modern parenthood.

I wrote a cohesive messaging system and a simplified way to look at early lactation so that the First 100 Hours approach would be accessible to anyone, whether they started from a place of significant medical background or none whatsoever.

As a deep dive into optimizing early lactation care practices, it enforces hard truths about infant feeding which have become lost in cultural and marketing messages about choices and guilt.

We have a moral imperative to ensure that families are entrusted with the knowledge they need to feed their babies without reliance upon products or equipment because the reality is that in today's environment, they may not always have access to those things.

Natural disasters, geopolitical conflicts, and other unexpected or unplanned events can disrupt the supply chain.

Power outages can render an electric breast pump useless and lead to the loss of frozen or refrigerated expressed human milk.

Infant formula recalls are not uncommon and can leave families desperate to find the formula their baby needs.

Creating dependence on formula instead of educating about how to breastfeed and express milk for feeding is irresponsible and unethical.

Promoting pumping over breastfeeding, or equating the two processes, is problematic and not aligned with the evidence base.

In fact, the arguments for pumping which say that it is the same as breastfeeding and which tell mothers that they "should not feel guilty" if they pump sound eerily similar to early marketing messages from formula companies about their products.

We have a duty to ensure that families understand that breastfeeding is equal parts immunological development, infection protection, and nutrition.

Viewing breastfeeding solely as a nutritional option ignores the fact that human milk is a live, biological fluid with lifelong impacts on the child.

Promoting exclusive pumping over breastfeeding minimizes the developmental importance of the interaction between mother and child as well as the physiological components of feeding at the breast.

A lens that views human milk as a set of macro- and micronutrients makes it seem easily replicable through formula.

Supporting informed choice in infant feeding can sometimes require difficult conversations about the differences between breastfeeding, exclusive pumping, and formula feeding, especially because many important outcomes are not visible or tangible in the short term.

However, this information is something that new and expectant mothers should encounter repeatedly and consistently from every healthcare worker in their universe.

Applying The First 100 Hours lens has the potential to make that a reality.

The First 100 Hours is a paradigm shift that helps you transform how you educate and support new parents.

Effective early lactation care requires the ability to remain focused on what can be affected or improved in the moment and what will need to be addressed going forward.

THE HEART OF THE FIRST 100 HOURS

I created this concept in 2014 while supporting a new mother in the hospital.

Her preterm twins were 36 hours old when she relayed to me that she had been essentially threatened by the twins' doctor in the NICU.

If she didn't bring a certain number of ounces of expressed milk for the babies within the next few hours, the babies were going to be given formula.

Despite telling the doctor and nurses that she had been faithfully following the pumping protocols and letting them know that she, too, was worried about not getting more than a few drops of milk, no help was given or offered to her.

They didn't call the lactation staff to assist her.

Instead, she was admonished for not bringing milk.

Fortunately, she asked her postpartum nurse how to get help with lactation.

After I provided the clinical lactation care she needed and then followed up with the NICU staff to keep them updated, I went home that day with this mother on my mind.

I really, really wanted to understand why the nurses caring for these less-than-48-hour-old babies, who were born too early for oral feeding and who were both receiving all the nutrition they needed intravenously, were doubting this mother's report that she was pumping according to the routine taught to her.

I also wanted to understand why a medical doctor, an expert

in preterm and sick babies, did not know enough about human lactation to have reasonable expectations of how much milk might be expressed in this time frame.

How could this happen?

The next morning, it dawned on me that our approach to educating the hospital staff was off.

They did not know how or when milk is made in relation to birth, so they had unrealistic expectations that were seriously impacting their patients.

There was a lack of trust in breastfeeding and human lactation on top of a lack of clinical knowledge.

These medical professionals operated with the vague notion that milk would suddenly "appear" a few days after a baby was born.

I knew that those giving birth in our hospital were not receiving accurate information about lactation in an organized, logical, simple way.

That was the day I outlined The First 100 Hours strategy and how it could make a difference.

The heart of this strategy is reframing the concept of "getting milk on the 3rd day."

Not only is it scientifically inaccurate, it is inherently confusing.

What is the 3rd day? If your baby is born on Wednesday, is Friday the 3rd day, or Saturday?

What if they were born at 10 pm on Wednesday?

That means they won't be 48 hours old until 10 pm on Friday.

If you think Friday is the 3rd day and you're wondering why there isn't copious milk production on Friday morning - when the baby is just 36 hours old - aren't you going to assume there's a problem?

To simplify the First 100 Hours, begin by drawing a timeline from the time of the baby's birth.

Include four sets of 24-hour periods, and make a heart after the 4th set.

That's your first 96 hours, plus 4 to get us to 100 and make it simple.

What should be happening in terms of lactation at that time?

What should we expect regarding how a baby is feeding at that time?

What should we expect in milk volumes expressed if the baby is not feeding at the breast?

What changes are expected in the breasts?

Milk is made beginning in early pregnancy; it's suppressed by hormones until the baby is born.

The earliest milk, called colostrum, is thicker than later milk because it is concentrated in nutrients, immune factors, growth factors, and more, and it contains less water.

Milk production shifts and increases within an average of 48-72 hours after birth.

This milk contains more water and all the nutrients and other components necessary for preventing infection and keeping a baby growing and thriving.

It's important to note that the 48-72 hour mark is an average, so it is not an exact predictor of when more milk will be available to the baby.

Also noteworthy: the feeling of fullness in the breasts has been shown to lag behind the actual increase in milk volume, so it is likely that more milk is being made before a mother can even feel it.

This is a commonly misunderstood time frame, and we see parents and healthcare workers alike struggling to set appropriate expectations.

Many research studies have been published that define and discuss "perceived low milk production" as a significant factor impacting the duration of breastfeeding, including exclusive breastfeeding.

However, in the earliest hours and days of lactation, when we ask mothers who are concerned about low milk production *why* they believe this, the answer is almost always related to their baby's behavior and feeding patterns.

It can also be because they are attempting to express their milk and seeing smaller volumes than they expected.

It is normal for new mothers to wonder if their babies are getting enough milk at the breast.

However, when their doubts or concerns about milk production and intake are multiplied by doubt and misunderstanding from medical providers, the situation becomes more complicated.

When a mother expresses concern over how much milk they are making and/or how much milk their baby is getting, they need accurate and actionable information and encouragement about how to tell the baby is getting enough milk as well as encouragement to continue to breastfeed, rather than uninformed and unfounded doubt on the part of healthcare workers.

Each misunderstanding of what is supposed to be happening compounds the problem and results in babies being fed formula they don't need, mothers being told to use breast pumps they don't need to use, and completely unnecessary stress and anxiety for new parents about breastfeeding "problems" that aren't there.

Early issues (like jaundice and hypoglycemia) which are multifactorial are too often mislabeled as breastfeeding problems that should be solved with formula.

New parents who are separated from their babies for medical reasons can be following everything they are taught about expressing milk by hand or with a breast pump in the First 100 Hours, but if they report that they are feeling discouraged about how much milk they are getting out, they may be offered a "sympathetic" way out (like "just get some extra sleep and don't worry about it too much") and a pat on the back that provides no real education or support that would help them sustain their milk expression routine.

It is normal for new parents to have doubts about breastfeeding as they are learning a new skill.

What is not normal or appropriate is for healthcare providers who should know the most fundamental science of human lactation to make the situation worse by building upon their worries and concerns with misguided information and advice.

When misunderstandings, misinformation, and misalignment of expectations lead to worried breastfeeding mothers feeding their babies formula, they also may stop breastfeeding.

Continuing to breastfeed even if a problem is suspected is a key component to ensuring that mothers retain the capacity to breastfeed in the long term.

The First 100 Hours approach is a reset button that provides the foundation of what to expect versus what truly constitutes a breastfeeding problem that requires intervention during the earliest hours of lactation.

HISTORY OF THE FIRST 100 HOURS APPROACH

I have been working in lactation support since 2001, starting as a volunteer in a peer breastfeeding support group and taking calls on a breastfeeding helpline.

In 2006, I started my first clinical work experience in a hospital under the mentorship of a skilled IBCLC.

At our hospital, there were typically three categories of patient feeding plans we would see in the morning census: exclusive formula feeding by choice, exclusive breastfeeding, and mixed/combination feeding (by personal choice or by medical necessity). It was pretty simple to figure out what to do that day: first, we would see the people asking for help, regardless of how they were feeding, then we would see the NICU moms, and finally, we would check in on everyone else.

Babies were not held skin-to-skin; all of the babies went to a "newborn nursery" away from their mothers for the first few hours regardless of mode of birth, and there were plenty of babies who couldn't keep warm, babies whose first feedings were with formula, and nearly universal pacifier use on the postpartum floor.

Many mothers who were combination feeding in postpartum assured us that they planned to exclusively breastfeed once they were home and their milk "came in," that they had done this with all their babies, and they'd breastfeed until they returned to work.

Occasionally, we'd get a call from one of them after they had gone home when they were engorged.

However, a little over-the-phone education, symptom management, and encouragement to get the baby on the breast usually did the trick.

The parents of babies who had somehow avoided formula in the newborn nursery and were now exclusively breastfeeding were always up for a visit by lactation staff even if they felt like things were going well.

If they needed help they definitely wanted to see us.

There were always a few combination-feeding dyads on the floor who felt that their plan to breastfeed had been interrupted.

These situations generally required most of our attention.

Many called us back after discharge, came in for outpatient, follow-up lactation care at their discretion, and joined our monthly breastfeeding support group.

Let me point out here one crucial factor that was missing, for better or worse: notice that I didn't mention pumping.

The only patients who were using breast pumps in the hospital were mothers of babies admitted to the NICU.

Pumps were not accessible, convenient, or affordable.

Locally, there was one retail baby store (open only during retail hours) that had 5 or 6 rental pumps available, but the cost was prohibitive for most families in town.

If a family was eligible for services from WIC*, they qualified to borrow (at no cost) an electric pump if their baby was in the NICU. If there was availability, some WIC-eligible mothers would get one to borrow when they returned to work.

(*WIC is the Women, Infants, and Children program in the U.S.; it is a federally-funded program whose goal is to ensure that all women, infants, and children have access to a healthy pregnancy and healthy food, including breastfeeding education and support)

You can easily see the gaps here: little to no accessibility to mothers without WIC eligibility whose babies were in the NICU, and no real "choice" to pump for anyone.

At the time, big-box stores weren't keen on stocking too many lactation products, and buying a good pump was really expensive if they even had one in stock.

Online ordering wasn't an option, either.

You might be thinking, "This was not that long ago!"

But the breastfeeding environment was really, really different back when health insurance plans did not cover breast pumps.

For example, as lactation care providers, we supported mothers who potentially could have used a breast pump as part of the resolution to their lactation issues, but since pumps were not available or accessible, we had to use other strategies.

If given a choice, I also believe that a significant number of people may have chosen to pump and feed their babies their own milk instead of formula.

Folks who felt uncomfortable or wary about breastfeeding for any reason would have had another option.

As lactation care providers, we did not have much opportunity to communicate with, counsel, educate, and learn from people who did NOT want to breastfeed but who still wanted to feed their own milk, and this left us with a gap in understanding.

Today, we are attempting to fill that gap with training on trauma-informed care, inclusive lactation care, and parent/family-centered care.

Our work at that time was mostly focused on people who *wanted* to feed at the breast (and often had taken prenatal classes to prepare for it) and had positive feelings about it, as well as those who *wanted* to breastfeed but might be experiencing challenges.

Essentially, lactation care providers on the whole served an overwhelmingly pro-breastfeeding population.

The more widespread application of the Baby Friendly Hospital Initiative and hospital policies that more closely follow the Ten Steps have changed the environment and helped to spread awareness of the importance of breastfeeding.

The Baby Friendly Hospital Initiative, which is guided by the scientific evidence collected and curated by the World Health Organization, offers a clear set of guidelines and criteria for how to support mothers and babies in the hospital with the information, assistance, and support they need to breastfeed.

Research has repeatedly shown that using the practices it recommends (even just a few) can significantly increase the hospital's breastfeeding rate.

Ironically, despite this more widespread awareness of what helps mothers initiate breastfeeding and the importance of breastfeeding, modern lactation care providers often encounter hostility, negativity, shame, anxiety, and clients who have been culturally prepared to "fight" with lactation care providers.

Product marketing and promotion messages have infiltrated the public's sentiment about breastfeeding, implying or stating outright that breastfeeding is just not that important in the grand scheme of life and that the challenges of breastfeeding are too much for most people to navigate, so they should just use a breast pump instead of "sacrificing" themselves for breastfeeding.

These commercial messages are amplified by social media influencers and others who stand to benefit - financially or otherwise - from demonizing breastfeeding and lactation advocates.

Healthcare providers who, themselves, struggled with breastfeeding or chose not to breastfeed use their platforms (in real life and online) to spread messages that will "save" others from the difficulties they faced.

The saying that "pumping is breastfeeding" is increasingly popular, this message tells exclusively pumping mothers that they "shouldn't feel guilty for pumping," a confusing message that plants the idea in mothers' minds that they should or should not feel a certain way about breastfeeding or pumping.

There seems to be more fear around the medical part of lactation (partly because the medical model still does not adequately study or care for the function of lactating people), and sometimes, we see more encouragement from physicians and other healthcare providers to rely on breast pumps than to breastfeed.

For lactation care providers, this means that we increasingly encounter mothers who are planning to use a breast pump immediately after birth, sometimes even to the exclusion of ever feeding at the breast.

We encounter clients with less educational preparation for breastfeeding because they have attempted to learn about breastfeeding from reading social media posts rather than from a structured prenatal course or training.

The total number of people who are breastfeeding has increased.

Because of their access to the Internet and social media, the amount of information they encounter in their lactation journeys has increased.

However, it can be difficult to filter and prioritize breastfeeding information on the Internet, so their overall knowledge may not be adequate or tailored to their specific needs.

Our clients have diverse needs and requirements, and our skillset as lactation care providers have doubled in order to accommodate the need to provide appropriate education and support for using breast pumps.

Though newly breastfeeding and pumping mothers alike need lactation care more than ever, they encounter messaging that tells them that hospital staff or lactation care providers will "pressure" them to breastfeed because they are somehow incentivized to promote breastfeeding.

Despite this alleged pressure, it is a lack of support and information that so often results in mothers stopping breastfeeding and/or pumping long before their intended goals have been achieved.

The landscape of breastfeeding care and support is drastically different from when I started volunteering in 2001, and we need to continue adapting to the changing needs of today's families.

Healthcare workers who are motivated to adapt to the modern breastfeeding environment should be prepared to:

- Take more feeding goals into consideration
- Provide more information about how to breastfeed
- Ensure that pregnant and lactating women have access to the help they need right when they need it

- Assist mothers with understanding lactation products and tools and their potential uses as well as harms

The First 100 Hours approach includes ways to support mothers with comprehensive information and support that will help them optimize breastfeeding and milk production right from the first hours after birth. This will allow them to feed at the breast for as long as they want and ensure their babies can thrive on human milk.

WHY WE SHOULD TRANSFORM THE MODERN EARLY BREASTFEEDING EXPERIENCE

T alk to almost anyone who has initiated breastfeeding recently, and they'll fill you in.

It's overwhelming, everyone tells them something different about how "worth it" it is, and they encounter conflicting information on a daily basis.

That is all on top of the inherent nature of how huge a change it is to go from being pregnant to having a newborn that you are responsible for 24/7 with essentially very little help from anyone you know, plus feeling like you are a different person all of a sudden.

Then, talk to any healthcare provider who works with new babies and their parents in an OB/GYN office, a hospital, a pediatric office, or a well-baby clinic. Ask them something about breastfeeding, and you'll probably get a different answer from each; some will be evidence-based and/or very helpful, while others will be much less so.

Breastfeeding in the U.S. and many other Western societies today can be confusing even in the best situations.

It starts very early.

Since acquiring knowledge about breastfeeding through a class or course during pregnancy has fallen out of favor, most people have to learn "on the job" when their baby is born.

We must change how we approach lactation education and support to address this new reality.

Here's the magic, though: the earliest hours of lactation are actually a good time to *experience* breastfeeding and milk expression.

The First 100 Hours after labor and birth are a time of recuperation and rest, a time for the mother to get to know her new baby, and a time to hear and internalize some simple messages about babies, breastfeeding, and making milk.

Envision your client - we'll call her Sophia and her baby Jonathan - spending the First 100 Hours of her baby's life with her newborn skin to skin most of the time.

Baby Jonathan tells them when he needs to be on the breast.

Sophia rests in comfortable, sustainable positions that allow her to latch her baby easily so he can stay on the breast for as long as he needs to.

Sophia is getting plenty of rest in a semi-reclined or lying-down position even when she's not asleep, and baby Jonathan is skin to skin or close enough to her that she can pick him up and respond to his communication signals.

Compassionate and helpful nurses, doctors, and hospital staff clearly explain what Sophia needs to know to keep breastfeeding when she gets home.

They talk about breastfeeding and newborn behavior in simple terms she can remember.

Sophia and Jonathan make a smooth transition from the hospital to their home, where they both settle in easily.

Sophia feels confident about what to do when she needs to put her baby somewhere safe so she can take care of her own bodily needs.

When she responds to her baby's communication signals, she feels connected to him, like she was meant for this job of being his mom.

She knows who to call if she has a question or concern about her own recovery or her baby's well-being.

Sophia believes that she knows what to do for Jonathan and herself.

The First 100 Hours is a very good time for this.

It's a perfect time to simply experience what breastfeeding, latching, and holding baby skin to skin actually feel like.

If a mother has chosen not to breastfeed, it's a perfect time to experience the rhythm and routine of expressing milk regularly around the clock, feeding their baby responsively using a method of their choosing, and holding their baby skin to skin.

In order to make this vision a reality for everyone who gives birth, we must:

- return to trusting breastfeeding and the biological process of lactation
- acknowledge that most babies latch and breastfeed well and that most babies do not need to have formula in the earliest hours of their life
- support and encourage new mothers to hold their babies skin to skin instead of swaddled tightly in layers in a crib or bassinet
- help new mothers experience the luxury and ease of being able to relax in comfortable positions while breastfeeding
- help mothers learn to massage their breasts and express their milk with their own hands
- remember that the answer to almost any breastfeeding problem is going to involve more breastfeeding
- reset breastfeeding as the default feeding method in general, with milk expression and alternate feeding methods as temporary interventions which allows us to provide additional education and support so that mothers can choose to continue them if they prefer or to resume breastfeeding if it has been interrupted

For those opting to exclusively express their milk for their babies, we must ensure that we provide within the First 100 Hours the routines, skills, and techniques to facilitate optimal milk production and minimal complications.

We must use the time we have in this critical period of a mother and baby's life to ensure that they have the information and support they need to recover from birth and get a healthy start together.

Unfortunately, healthcare professionals are offering vastly outdated information about human lactation or allowing their own experiences to dictate what they tell people who are trying to breastfeed and make milk.

There are too many people out there who are invested in old or inaccurate beliefs about breastfeeding or using words and terms that minimize it.

The falsehoods they convey, their skepticism, and their doubting comments are causing harm.

Too often, new parents relay stories of things that were said to them in the First 100 Hours which made it feel as if they needed to choose between breastfeeding and health (their baby's or their own.)

We can transform this.

We can promote breastfeeding and human milk so that mothers and babies thrive.

We can demonstrate that breastfeeding works, it is reliable, and if problems arise, help is available.

Fearing negative outcomes from exclusive breastfeeding is not necessary because safety net protocols are already in place in hospitals to catch issues like low blood sugar in newborns, jaundice, and early excessive weight loss.

Doubting breastfeeding does not make babies safer.

Confidence in breastfeeding comes from acquiring experience with the act of breastfeeding.

We can change things for future families by developing consistency in education, building (or rebuilding) trust in breastfeeding, and fostering our collective ability to facilitate the healthcare field as a whole to support breastfeeding and lactation. It begins with The First 100 Hours approach.

VISION, MISSION, AND OBJECTIVES

Vision: New mothers and babies thrive in the First 100 Hours after birth

Mission: By providing accurate, evidence-based information about human lactation and infant development, new parents are guided with appropriate expectations about newborns and supported with the skills and techniques they need to optimize breastfeeding and health within the First 100 Hours and beyond

Objectives:

- Help new mothers establish appropriate expectations for how to meet their breastfeeding goals
- Increase the rate of mothers breastfeeding their newborns exclusively from birth
- Decrease the incidence of non-medically necessary formula supplementation in the hospital and beyond
- Increase the duration of mothers feeding their infants at the breast

Encompassing the World Health Organization's Ten Steps, the protocols of the Academy of Breastfeeding Medicine, and adhering to evidence-based best practices accepted around the world, The First 100 Hours is an approach to early lactation care that embodies all of these recommendations and expands upon them. By including the information and support needed to establish optimal milk production without latching a baby, we can better prepare new mothers who are separated from their newborns, those who are struggling to latch their babies, and those who have opted to express milk and use bottles to feed their babies.

Mothers who opt not to or are unable to lactate are fully informed and educated about developmentally critical behaviors and practices around infant feeding to optimize their and their babies' First 100 Hours of life when feeding donor human milk and infant formula.

This approach allows us to address all feeding options and potential obstacles while preserving breastfeeding, milk production, and health.

Using The First 100 Hours Evaluation Form (see Appendix F) allows a lactation support provider to conduct an organized evaluation of the current lactation history of their client and baby and determine what needs to happen next, whether breastfeeding is going well or it has become complicated.

It brings the most critical factors into focus.

It offers a clear reminder of the specific education and techniques that still need to be provided or reinforced or if further referrals need to be made for additional types of lactation or other health care.

This approach is designed to be used in the hospital, after discharge, at home, and the first pediatric or clinic visit, helping to create stronger threads of continuity in messaging, support, and education.

It can also be applied after the First 100 Hours have come and gone.

Through an understanding of what happened during that time, the lactation support provider has an optimal picture of the scope and history of breastfeeding and lactation since birth.

This can still be helpful when assessing a baby who is already eight days, two weeks, or even 3-4 weeks old.

What happens in the First 100 Hours, just like what happened during labor and birth, continues to have an impact on lactation, sometimes for weeks or more.

The First 100 Hours approach promotes and facilitates respectful,

family-centered care of the dyad.
This is what preserving the dignity of new families looks like.
This is how we transform the modern breastfeeding experience.

The First 100 Hours approach is designed to:

- help mothers and babies achieve success with breastfeeding and lactation in the future
- increase the potential that they will breastfeed and feed human milk longer
- prevent or minimize complications of lactation
- establish a level of milk production which is synchronized with the baby's needs
- engage all tiers of lactation support and healthcare workers to facilitate breastfeeding and human milk feeding
- optimize infant immunity
- optimize infant nutritional foundations
- optimize infant development
- optimize the infant microbiome
- optimize hormones for the infant and the mother
- optimize the infant's absorption of nutrients
- avoid potentially introducing pathogens to the infant's gut
- avoid the introduction of foreign (non-human) proteins to the gut
- allow proper closure of the infant's gut lining
- avoid overfeeding of the infant
- facilitate early recognition of red flags for breastfeeding and milk production problems

RESTORING YOUR FAITH IN THE PROCESS OF BREASTFEEDING

There are many reasons you might find yourself disagreeing or struggling with what you are reading in this book.

You may have personally breastfed and faced challenges and/or little to no actual lactation support.

You may have watched someone you love struggle with breastfeeding or milk production.

You may not have children and feel unqualified to have an opinion about breastfeeding.

You may have had experiences with babies who had poor outcomes while breastfeeding.

You may have been exposed to the three themes of the mainstream American journalistic playbook regarding reporting about breastfeeding: the author inserting their personal baggage into an article that is supposedly objective or neutral, inaccurate claims that the science is "wrong" or lacking; and the message that lactation care providers are "bad/rude/pushy," etc.

There may be other factors that make this information challenging for you.

What you've seen happen with breastfeeding may appear to be today's "norm" to you, but it is not the biological norm.

A responsive breastfeeding relationship that synchronizes the needs of the developing child with their mother's needs for recovery from birth and energy to provide their care is the biological norm for our species.

The biological norm is a simple course of breastfeeding with minimal interruption or interference and the mother's constant presence as primary caretaker.

The biological norm is exclusive breastfeeding for the first 180 days of an infant's life, followed by continued breastfeeding with the addition of complementary nutritious, culturally appropriate family foods.

A natural, mutually reinforcing weaning process that occurs when both mother and baby are developmentally ready, usually after several years of breastfeeding, is the biological norm.

Here's your moment of truth: From about 2015 to 2020, I struggled with continuing to believe in these biological norms.

I saw the rise of breast pumps, the repeated separation of mothers from their babies throughout their infancy due to the realities of modern society, and the gradual easing of breastfeeding promotion as it gave way to promoting the prioritization of feeding human milk.

I was seeing a shift in the mothers who were attending our Baby Café, and though some were falling into the biological norms because it seemed easy to them, far more were struggling mightily with every fiber and thread of the whole cloth that breastfeeding is.

Far more mothers were facing complexities in their breastfeeding journeys.

Far more mothers were comparing themselves to other mothers on social media.

Far more mothers were facing layer upon layer of structural barriers to breastfeeding and finding that they needed to give something up just to feel okay and make it through the day.

But then our whole world shifted, and we all stayed inside our homes, trapped by our collective lack of understanding of a virus that seemed to be taking over the world.

Those mothers I had gotten to know at Baby Café weren't coming in to see us, but they also weren't having to leave their babies in someone else's care while they went to work.

They were frustrated, annoyed, scared, upset, depressed - but not about breastfeeding.

You see, when they found themselves at home with their babies, they stopped pumping as much and started breastfeeding more.

Many were still working at home from their laptops, but since they didn't have to leave their babies, they didn't need or want to use their breast pumps.

They surprised themselves with how much they enjoyed breastfeeding.

They reached out to tell me and the rest of our IBCLC staff how much they realized they had been missing.

I listened intently to everything I was hearing from mothers in the pandemic, and one theme emerged: everything else felt out of control in their lives, but breastfeeding didn't feel as hard anymore.

I realized that I personally had been equivocating on talking about breastfeeding as the biological norm because I had been working with so many mothers whose primary experience with lactation was pumping.

I knew then that I had to write this book and I had to really get serious about helping others to learn how to work through their doubts and concerns about breastfeeding promotion.

I fully understand that it is not easy to take a stand about breastfeeding.

It can seem very controversial and difficult to discuss, and encouraging or promoting it may even seem to go against your own personal experiences with breastfeeding.

I am willing to walk with you to help you find out how you can become an advocate for breastfeeding.

I know you can do this because you have already committed to being a healthcare provider, which means advocating for biological norms and healthier outcomes.

Here's what it will take for this book to make a difference for you and those you serve:

- you will need to put your own experiences (or lack thereof) with breastfeeding in a box on a shelf when you are helping others
- you will need to remind yourself frequently of the context of breastfeeding and what it means to be a mammal
- you will need to internalize that breastfeeding is important to other people
- you will need to set aside how you personally feel about breastfeeding in order to provide lactation care

This is huge.

It means that if you had a wonderful, beautiful experience, you'll need to set that aside, and if you had a horrible experience that you would never wish on anyone, you'll need to set that aside, too.

Your lactation experiences and opinions are very important - but they are part of you and your life.

They are not part of your client's life unless you bring that suitcase into the consultation room.

The people you are responsible for helping with breastfeeding have their own suitcases: their own goals and plans, and their own needs.

They do not need you to cover them up by spilling out your own.

I won't sugarcoat it: this can be hard and happen in subtle ways you don't even recognize.

Your personal biases about breastfeeding, which come from your own experiences around it, may seem useful, logical, or helpful to you.

You likely talk about them with patients, clients, and people you know in your personal life.

Perhaps you feel that a lactation product or tool "saved" your breastfeeding relationship.

Maybe you felt like your lactation care provider, nurse, or doctor didn't support you so you want to "warn" others against asking for help with breastfeeding.

Or it could be that you had a lovely breastfeeding journey and you just want everyone else to have a magical time, too.

All of these can be completely off-base and out of context for the other person.

Talking about your wonderful experiences might make her feel like she is failing because she does not feel confident or great about breastfeeding.

Your cautions against asking for help may be 100% wrong for your cousin, who really needs clinical help.

Your experience using that lactation product may have been the perfect intervention for your situation, but completely unnecessary or harmful to your client.

Unfortunately, your personal biases about breastfeeding can leak out in the lactation care you provide, too.

You may find that you steer most of your patients toward pumping, "just in case" they struggle with milk production.

You may look for latch issues when the mother reports no problems and the baby's output is normal.

You may realize that even though you say you support breastfeeding, you end up telling nearly all of your patients to supplement with a little bit of formula.

Maybe you've seen so many people struggle to succeed with breastfeeding that it feels like success isn't possible.

In order to restore your faith in breastfeeding, you will need to understand it better.

You will need to reset your own expectations about the First 100 Hours and how experiences during that time frame can establish patterns—for better or worse—that persist throughout a mother's breastfeeding journey.

You will need to train your brain to consistently return to the basics of early lactation rather than jumping to conclusions about what is happening.

Parents are hearing plenty of voices and cultural messages saying "It's ok not to breastfeed."

If you are part of the healthcare team, your responsibility and duty to do no harm means you must be a voice that advocates FOR breastfeeding.

It is not your responsibility to "save" people from breastfeeding, even if breastfeeding was hard for you or if you believe stopping breastfeeding was the right choice for you.

It's ok to advocate for breastfeeding even if YOU didn't do it.

It's not hypocritical - we do this in all types of situations in life.

We are all different, and we find ourselves with different experiences.

We connect by supporting others who are on their own journey.

Breastfeeding can be hard for some people, and there's no way to predict who will experience challenges.

Acknowledging that breastfeeding is hard isn't the same as saying it shouldn't be done.

As humans, we do plenty of hard things.

Breastfeeding is possible, and people do it all the time around the world.

That is the real context of breastfeeding.

If we truly want to make a difference and evolve the modern breastfeeding experience, starting with the First 100 Hours, we need to facilitate the *process* of breastfeeding for every dyad before we ever consider introducing *products*.

Teach people how to breastfeed.

Teach them how to sustain it without products.

That is ethical, sustainable lactation care.

Access to feeding products like formula and lactation tools and devices is never guaranteed.

For a mother who doesn't know how to get milk out of their breasts without a breast pump, a power outage or forgotten pump parts will make pumping impossible and force them to figure out how to remove milk.

For a mother who has not been supported to breastfeed, a formula recall can make life extremely challenging.

Encouraging reliance on lactation tools and feeding products creates an unnecessary vulnerability and dependence for mothers, whom we so frequently say that we wish to empower.

Understanding mothers' choices to use lactation tools and feeding products beyond the First 100 Hours is outside of the scope of this book. Still, when supporting mothers before and during the First 100 Hours, it is reasonable to say that it is incredibly important to teach the process of breastfeeding and avoid using products as much as we can.

We must all adopt the mindset that a new mother will usually not need a tool or formula, especially in the earliest hours of lactation. With a well-defined plan for continuity of care (from hospital to home to community), breastfeeding can be trusted and supported.

RESETTING EXPECTATIONS

O ur teaching before and during the First 100 Hours must revolve around appropriate newborn expectations.

When we consider early breastfeeding outcomes, we need to be able to determine which outcomes are symptoms of a breastfeeding problem versus which are outcomes of a poor lactation management routine.

For example, let's examine a case in which a 36-hour-old baby has had only eight feedings since birth.

It's critical to know that we expect a baby to have fed eight times by the time they are 24 hours old, so we are looking for potential reasons that the baby may not be feeding as frequently as expected.

Upon further conversation with the baby's parents in their room, we observe the baby swaddled tightly and sleeping in his crib.

The parents proudly report that he's been very calm and quiet since they learned how to swaddle him from their last nurse.

Is the lack of sufficient feedings because breastfeeding isn't working or is it because there isn't enough milk, or is it more likely that the baby is not being kept skin to skin and the tight swaddle is interfering with his cues to go to breast?

Is the lack of sufficient feedings a symptom of poor breastfeeding, or is it a result of poor information on newborn care provided to the parents?

What is the optimal approach to resolving this situation?

It is easy to see how giving the baby some pasteurized donor human milk or formula at this point seems like a "safety" measure; giving the baby food appears on the surface to "solve" this breastfeeding problem.

The problem with this solution is that it only answers one of the dyad's needs, leaving the others unmet, including the baby's need to be skin to skin, the baby's need to suckle at the breast to get milk, and to meet his suckling needs, the mother's need for milk removal from her breasts to signal her brain to make more milk, and the mother's sense of security in her body's ability to meet her baby's needs.

Conversely, suppose we encourage the baby's mother to unwrap him and put him skin to skin so that he can begin the sequence of movements that initiate breastfeeding.

In that case, we teach her how to respond to a breastfeeding problem with the simplest and gentlest intervention.

We can also teach her how to hand express her milk and feed it to him with a spoon or cup to increase his intake before he latches and nurses.

There are so many steps between recognizing a potential breastfeeding problem and actually needing to circumvent breastfeeding by feeding formula.

Many early lactation misunderstandings stem from inaccurate timelines of when events are expected to happen during the First 100 Hours post-birth.

Let's look at the optimal early breastfeeding experience: a healthy, term baby is born and placed skin to skin immediately.

Within the first hour or so, he latches and feeds. He is kept skin to skin most of the time and cues to go to the breast often.

When there, he latches and nurses without causing pain or nipple damage.

Some breastfeeding sessions are longer than others, and some intervals between breastfeeding sessions are longer than others.

Cluster feeding, in which the baby nurses multiple times within a few-hour period, then takes a break for a few hours and is calm and/or sleeping, is normal and expected.

This nursing pattern at varying intervals and for varying

amounts of time is also normal and expected.

Baby's diapers are counted to observe his output: in the first 24 hours, he should have at least one wet diaper and one or more stools (which are black and very sticky); in the second 24 hours he should have at least two wet diapers and two or more stools (usually still black and possibly turning to green); in the third 24 hours he should have at least three wet diapers and three or more stools (black or green), and in the fourth 24 hours, he should have at least four wet diapers and three or more stools (which are greenish to yellow and seedy by this time).

In the First 100 Hours, diapers are tracked because they indicate not only that breastfeeding is progressing, but that other important bodily processes are occurring as expected.

In the First 100 Hours, diaper output is NOT a direct reflection of the intake of milk.

The progression of color of the stools is a particularly important indicator of intake in breastfed babies.

When copious milk production is observed and the breasts feel fuller, diaper output transitions to being a strong indicator of sufficient milk intake.

Therefore, after the onset of copious milk production, the baby should have 5-6 wet diapers per 24 hours, and they should be yellow and seedy in exclusively breastfed babies.

This is one of the areas that is most frequently misinterpreted.

When a baby is assumed to be not *getting* enough milk, generally the assumption is that the baby's mother isn't *making* enough milk, when it could just as easily be that the baby isn't feeding frequently enough, latching effectively, or suckling well at the breast.

New mothers and their partners and other family members may worry that the baby is feeding too often or too long, having feedings too close together, or not having enough diapers.

These seem like logical conclusions if their expectations are that a baby will feed for a set amount of time at regular intervals and immediately have diapers that need to be changed constantly as soon as they are born.

Babies aren't pre-programmed to eat or nurse at specific times or specific intervals.

In fact, they are following internal signals that lead them to communicate outwardly, through cues, that they need to go to the breast.

Sometimes, they will show cues again not long after leaving the breast, and other times, they will have longer gaps between nursing.

This variation is directed by the baby's needs, which only the baby can communicate.

Their need to go to the breast and nurse is not dictated solely by their nutritional needs; there are many other reasons babies need to go to the breast.

The timing of when an individual baby will need to go to the breast is not predictable.

We can't use a clock or a schedule to determine when to feed a baby or how long they should nurse.

Therefore, it is the mother's role to respond to her baby's communication signals by bringing him to the breast when he asks and keeping him there as long as he is actively nursing.

When done nursing for that session, he will be content even if he is moved away from the breast.

Though he may still need to be held, he won't put his fists to his mouth or root toward the breast.

(If he is, he is communicating that he still needs to be on the breast.)

Other common practices can interfere with the baby's ability to communicate that he needs to go to the breast.

For example, if he's being tightly swaddled and placed in a bassinet or crib, he may not be able to use his hands to signal that he is ready to go to the breast.

If he is using a pacifier, it may meet his suckling needs, which are meant to be met by suckling at the breast.

If he is not being kept skin to skin, he may have to re-orient himself and his movements before being able to latch and nurse effectively, wasting precious time and energy.

Infants learn during their feeding opportunities and experiences. They are learning whether their communication signals will be heard and respected.

They flourish when they're afforded the appropriate human contact that their brains require during feeding times.

The process of learning the rhythm of this communication between baby and caregiver is facilitated through the practice of "rooming-in," where babies and mothers are kept in the same room after birth, rather than moving the baby back and forth from a nursery.

It is a synchronization of need and response, or what is called responsive feeding.

Babies communicate their needs, and through the response of their mother bringing them to the breast, a synchronization of milk production simultaneously occurs.

It is a circle of feedback that requires each to respond to the other. Interference (through practices like limiting feeds, swaddling, pacifiers, and feeding expressed milk or formula away from the breast) results in different outcomes.

By helping parents understand that the first 100 hours require only a few simple goals, we can help them manage this relatively brief period of their baby's life.

We expect the mother and baby to be skin to skin most of the time, for the baby to communicate that he needs to go to the breast frequently, that those communication signals will be met with the response of being brought to the breast and facilitated to nurse for as long as he likes, and that the mother and baby will rest in between.

Frequent breastfeeding is a feature of newborn babies, not a signal that anything is wrong.

SIMPLE, GENTLE SOLUTIONS

Nearly every problem that can be encountered while initiating breastfeeding in the First 100 Hours of life can be addressed with a gentle and simple solution.

Skin to skin and hand expression are the primary tools for breastfeeding and lactation concerns, and they are free.

When adopting Baby Friendly's Ten Steps or similar practices in a hospital setting, it is common to see that the staff learns to use these strategies the first time a problem arises, but in time, they may lose faith in them or fail to use them frequently enough.

When healthcare workers don't trust that skin to skin and hand expressions will help, it's unreasonable to expect new parents to trust these techniques.

If a baby is having trouble latching or is acting sleepy and uninterested in latching, skin-to-skin contact is the first step in resolving the problem.

If the baby needs milk, hand expression can be accomplished while the baby is skin to skin, and the milk can be fed with little interruption.

The baby may then continue to rest or, having taken in some calories, begin to arouse and seek to latch again.

Whatever the outcome, if this problem arises again, the response must be the same: bring the baby back into skin to skin contact, attempt to latch, and hand express milk.

Observe the baby for signs of adequate intake and intervene only as medically indicated.

Suppose the baby requires additional intake and donor milk or formula is given.

In that case, the first steps are STILL to maintain the baby in skin to skin, hand express and feed the baby the expressed milk, and then, and only then, if additional volume is not required and milk can no longer be expressed at that moment, feed the donor milk or infant formula.

If it happens again, the steps are the same.

Breastfeeding, skin-to-skin contact, and hand expression are ALWAYS indicated for babies who are not taking in enough milk or who cannot latch or breastfeed well.

Should a situation arise where it is medically indicated to offer the baby pasteurized donor human milk or formula, the cycle of skin to skin, breastfeeding, and hand expression must continue as well.

These strategies protect the baby, their mother, and the milk production process.

If they are skipped, lactation can be compromised.

Continuing skin to skin, breastfeeding, and hand expression even when donor milk or formula are needed helps to establish that mothers should continue them anytime during lactation that they encounter a problem.

When we protect milk production early and continuously, we prevent low milk production from compounding other complex situations.

Here are some more specific teaching strategies for helping new mothers utilize their breastfeeding skills when facing common early breastfeeding challenges.

Skin To Skin

Skin to skin ensures that the mother is resting and connected with their baby.

For mothers who plan to exclusively express their milk for their baby, this is a crucial way to ensure they can optimize their milk production right from the beginning while offering them an ideal way to remain connected with their baby physically.

If feeding from the breast is not desired, the baby can be gently redirected and moved into a safe position for feeding from a bottle or cup.

Skin-to-skin contact facilitates the baby's maintenance of normal physiological stability, prevents the baby from wasting calories trying to stay warm, and offers quick access to the breast.

Why would we ever skip this step?

This step is a reset button for a baby and mother to synchronize with each other.

Holding her baby in only a diaper against her bare chest, the mother can be secure in seeing her baby find comfort in the place that feels closest to where they have been growing for nine months.

Skin to skin is also the simplest form of bodywork.

It resets just about every key physiological process for all babies.

For late preterm and preterm infants, it puts them in a physical state where true growth and development can happen because it removes so many other stressors that use up valuable resources for them.

Throughout the whole journey of breastfeeding and lactation, we can find ourselves talking to parents about options for professional bodywork services and allied health providers to improve their baby's feeding capacity - but this one does not require them to go anywhere, make any appointments, or spend any money or insurance benefits.

Bodywork of all kinds can be incredibly helpful for babies, and skin to skin is bodywork, too.

All that crawling, stretching, neck craning to see their parent or caregiver's face and eyes, stretching their arms and pushing up, moving their legs to adjust their position, scrunching up, and then stretching back out—not to mention the positional stability they can experience while skin to skin.

It can seem vastly different from the common breastfeeding positions mothers see represented visually on social media and in books, TV, and movies.

We have the opportunity to remind parents of this physiologically normal positioning (which is counter to so many positions babies are placed in today, e.g., back sleeping for safety, car seats to protect them in vehicles, swings or other containers for movement and stimulation or calming).

When parents are stressed, overwhelmed, or concerned about their baby, they may not realize that there are free or low-cost interventions that are very effective; instead, they may assume that their baby needs a type of therapy or a product (device, medication, etc.) to resolve the problem.

Skin to skin is an amazing bodywork tool.

When it is practiced by a mother or other caregiver who is encouraged to approach it with the intention of calming and relaxing both themself and the baby, it can be a useful tool for them to rest and relax as well.

Skin to skin only requires simple education that can be provided before birth or even just after birth.

For maximum safety, it's critical to ensure that when mothers hold their babies skin to skin in the earliest hours after birth, they are observed by birth attendants and taught to ensure that the baby's face is turned to the side and free to move and that the baby is held in between and with their head slightly above the breasts.

Mothers and other caregivers who are being shown how to hold their baby skin to skin can be reminded not to fall asleep with the baby skin to skin when they are alone or in positions where the baby could fall.

The Latchbetter Framework

It is fascinating to observe that when a baby is latching well, no one seeks or seems to need any details about it.

A mother whose baby is latching well does not need to analyze it; she understands intuitively that it is working.

However, when a baby is not latching well (causing pain, creating physical trauma to the nipple, suckling only faintly, suckling with a slow or disorganized rhythm, etc.), details become necessary -

and mothers know this instinctively as well.

It feels disruptive to her in a way that drives her to take action: to seek help, to ask questions, or to stop breastfeeding.

It would be easy to say here that if the baby is latching well and there are no complaints, then there's no need to be concerned.

In fact, that is what many earlier writings have taught, and many lactation care providers have been taught this over the years.

In order to evolve and modernize the American breastfeeding experience, though, we must address this differently.

Today, expectant parents are exposed to so much messaging around what to expect with breastfeeding that we are responsible for helping them sift through everything they've heard and filter out what will actually be useful to them.

The modern cultural narrative tells them that breastfeeding will be hard, painful, and something to "endure."

They may be interpreting that to mean that latching should be painful or that there's no way to breastfeed without pain.

Our role, then, is to teach new mothers how to latch their babies and what to watch out for other than pain.

We can explain five main areas to help you envision what effective, functional latching looks like.

We may not have the opportunity to describe every element listed here to every new mother we see, but this framework can inform how you look at a baby's latch and separate out its details.

State: the dyad begins a nursing session in a calm state, both breathing normally without difficulty, both focused on the task at hand with no big distractions or other responsibilities, at feeding sessions directed by the baby's communication signals

Positioning: both members of the dyad are settled and comfortable from the moment they begin nursing until the moment they are done; the baby's body is fitted on or against its mother's body; both mother and baby feel that their bodies are stable so that they don't have dangling limbs or tense muscles trying to hold them in place; the baby is brought close to and facing the breast with their nose lined up with mother's nipple and their body aligned in a neutral position; the baby has space enough to lift their chin toward the breast rather than tucking it toward their own chest

Anatomy: the baby has normal anatomical features in its oral cavity, including the range of motion in the jaw to gape widely and support the weight of the breast while nursing: the ability of the tongue to lift anteriorly and posteriorly, and cup the nipple to form a channel for liquids; enough padding in the cheeks to fully create a closed chamber within the mouth during suck; an intact swallow reflex; and the mother's nipples evert when stimulated by the baby's latch

Movement: as the baby suckles, their jaw glides easily, they suck rhythmically with periodic pauses for extra breaths without detaching from the breast, and they nurse actively for at least several minutes without any problems managing the changing flow of milk from the breast throughout the nursing session

Outcomes: upon detaching from the breast, the mother's nipple appears normal and slightly elongated with no skin damage, the dyad is relaxed and content, the baby acts satiated, and the feeding falls within the context of an overall breastfeeding situation, which is demonstrating normal outcomes in terms of frequency of feeding, diaper output, weight changes, and overall infant behavior.

With this framework, it is easier to provide a more nuanced picture of effective latching and to help determine what to do if something does not seem right.

It is especially important to ensure that emphasis is placed on the fact that pain is never a normal part of the lactation process.

Pain is a signal to the brain that tissue damage is already occurring or could potentially result from the current situation.

The brain reacts accordingly, interfering with the milk ejection reflex, which results in less milk being output to the baby (or to a breast pump).

A functional latch works BECAUSE there is no pain, not in spite of pain.

Creating opportunities to discuss effective latching allows new mothers and parents to explore what they have heard, what they expect, and what they can watch out for as breastfeeding progresses.

Hand Expression

In the First 100 Hours, our goal is to set people up with a foundation of the skills and techniques they need to feed their babies.

We teach them how to know when the baby is communicating that they need to go to the breast, we teach them how to latch the baby, and we make sure they have plenty of opportunities to experience what latching and nursing feel like.

We give them the information they need to ensure their baby's progress with breastfeeding over the first few days.

A critical component of this progress is learning how to hand express their milk.

When people know how to hand express their milk, it gives them confidence that they are really making milk, and it allows their babies to get their milk for a longer period.

If they stop breastfeeding for any reason at any time, they still have the option of feeding their milk because they know how to express it even if they have never used or do not have access to a breast pump.

All people who have given birth need to be taught hand expression.

Every person who gives birth deserves to be taught to hand express so that they can be in control and in charge of what's happening with their body.

Whether their baby is in their arms, in the NICU, separated from them for legal or safety reasons, unable to be breastfed because of a contraindication to breastfeeding, or, sadly, unable to survive, every birthing mother deserves to know how to manage the approaching moments and hours of the onset of copious milk production.

Physiologically, for a person who has given birth, somewhere between 48 and 72 hours post birth on average, there will likely be a surge in milk production.

Sometime after the surge happens, the sensation of breast fullness will occur.

In the absence of frequent and effective milk removal, engorgement of the breasts can occur, so in cases where we are aware that milk is not being removed or will not be removed, we must ensure that there is action to prevent engorgement.

Many factors impact when the surge of milk production occurs and how much milk is made; we simply do not currently have a way of predicting the timing or volume of milk.

Therefore, it is not possible to "diagnose" delayed lactation or low milk production before 72-96 hours at the earliest, especially when a mother is exclusively breastfeeding.

Hand expression during these earliest hours provides the needed oxytocin stimulation in the breasts and the milk removal signals needed by the brain to promote milk production.

It also usually results in some expressed milk being made available to feed the baby away from the breast.

When milk production needs to be continued and maintained for a baby who is or will be breastfeeding, education and techniques should be provided to protect adequate and appropriate milk production.

When milk production is to be suppressed for any reason, education about hand expression can allow small amounts of milk to be removed only to a level of comfort and generally prevent engorgement.

Oxytocin is one of the primary hormonal drivers of the lactation process, so it is important to consider its role in context when discussing hand expression versus using a breast pump.

A good friend of mine who is a NICU nurse and IBCLC told me this story from her experiences working in the hospital.

One evening, she was talking to a couple at their baby's bedside in the NICU about 48 hours after the baby's birth.

They had been set up with a breast pump, discharged home from the postpartum unit with all the instructions for pumping, and then arrived for this visit without any expressed milk for the baby. They were very frustrated about that.

They said, "We're doing everything you told us to do. We're following all the rules, and there's just no milk. Nothing is happening."

My calm and patient colleague reassured them, saying, "Well, since you're here, interacting with your baby, let's get some hand expression going."

The father said, "What are you talking about? We spent hundreds of dollars on this breast pump. Shouldn't we be pumping?"

My friend replied, "Well, your hands are attached to your brain! Hand expression will send some different signals to the brain than the pump. The pump is only attached to the wall. How about if we try both hand expression and then some pumping?"

When my friend told me this story, I thought it was absolutely brilliant. What an interesting way to examine the difference between what happens during hand expression and when we plug a machine into the wall and ask it to interact with our body.

Every opportunity we have to talk with new parents, we can encourage them to increase oxytocin by holding their baby skin to skin, breastfeeding frequently, and hand-expressing.

The removal of milk from the breasts also increases prolactin levels, a key factor in sustaining milk production over time.

Every time the baby is suckling, every time the baby is skin to skin, and every time we hand express drops of milk, prolactin levels rise.

The frequency of prolactin peaks is one of the things that drives milk production.

If long gaps are occurring in between the peaks, milk production decreases.

Frequent feeding (8 or more feeds every 24-hour period) is so important in early breastfeeding.

This is why we need to encourage hand expression whenever a baby isn't going to the breast—the total number of milk removals (by baby suckling, hand expression, or, later, pumping) is ideally at least 8 times per 24 hours.

Though a common cultural narrative claims that many people cannot or do not make enough milk, as previously discussed, primary lactation insufficiency cannot be determined in the First 100 Hours of a baby's life.

This is critical because when we observe that a baby is not having the expected outcomes in terms of diaper output, weight, or jaundice, it does not mean that its mother is not making enough milk.

It MIGHT mean that, but we wouldn't know that at this time.

Instead, we must first act on the knowledge that the baby isn't getting enough milk (which could also be because of something the baby is doing or is not doing correctly).

Resolving the problem always begins with improving the breastfeeding technique and increasing milk intake in the simplest, gentlest way possible.

We do know, however, that poor early lactation management routines can also result in low milk production, or what is called secondary low milk production.

For most people, milk production is something they can control by following the fundamentals of what we know about how and when milk is made.

Educating parents about appropriate expectations for breastfeeding right from the first hours after birth optimizes their capacity for milk production. Conversely, if they misread what is happening, they may develop unfounded doubts and concerns about how much milk they are making.

I can very clearly recall a case in the hospital that illustrates this.

I was consulting with a new mother on the morning of her and her baby's discharge from the hospital.

Breastfeeding was going very well, and all indicators were excellent: no nipple pain or trouble latching, normal diaper output, minimal weight loss, and no jaundice.

She relayed to me that the baby had breastfed well and frequently throughout the night, so she was tired but content.

We talked about cluster feeding and why it is normal for babies sometimes to have more frequent feedings while feedings are more spread out at other times.

While I was still there, her baby's pediatrician came in to do a final check and prepare the baby for discharge.

He asked her how breastfeeding was going, and she told him about the frequent feedings.

He became concerned.

"Well," he said, "it might be best to give the baby a little formula after each feed. It sounds like maybe he isn't getting enough from breastfeeding."

When he left, I picked my jaw up off the floor and asked her how she felt about that. Unfortunately, the damage was done.

She now had seeds of doubts about her baby getting enough milk.

I reinforced the information I and her nurses had given her about how to tell if her baby was getting enough, but it was already too late.

An exclusively breastfed baby was now going to have formula because of inaccurate information about what is normal for newborn babies.

We cannot assume that a baby's breastfeeding problem in the First 100 Hours is caused by their mother not making enough milk.

That is often not the problem at all, and if we begin with that assumption, we may miss the fact that the baby isn't latching or suckling well.

In other words, it's not only about milk production.

If we assumed the issue was always that the baby was not getting enough milk at the breast and we filled that gap by feeding formula, we would miss many other types of breastfeeding problems.

Babies have their own role in breastfeeding success as well.

If we suspect or know that a baby is not latching or suckling well, hand expression is the simple, gentle solution that fills the gap to keep them fed AND protect milk production while maintaining breastfeeding.

This also helps to avoid giving the impression that formula "fixes" a breastfeeding problem; instead, reinforces the education about normal newborn breastfeeding patterns and encourages skin to skin and hand expression as the tools we use to keep breastfeeding on track, whether or not formula is medically needed.

Importantly, during the First 100 Hours, ensure that hand expression is not promoted as a way to prove that there's milk in the breast, that the baby is getting milk when they're breastfeeding, or that a large volume of milk is being made.

We must be clear about what is normal to expect from early hand expression.

We must explain that what we might see is drops, possibly just enough to collect onto a spoon.

During early hand expression sessions, a few drops may glisten on the nipple.

That does not mean it is not working.

These small volumes often cause people - mothers and healthcare workers alike - to say, this doesn't work and feels like a waste of my time.

It is normal to see drops of colostrum glistening on the nipple, colostrum dripping from the nipple, and colostrum spraying out of the nipple.

It is not possible to predict what we will see from any one mother at any given time.

The procedure of hand expression is the most important thing that is occurring.

Colostrum (a name we give to the earliest milk) is a very concentrated form of human milk that is produced in very small volumes to meet a baby's limited capacity to manage liquid flowing into their mouth and give them a little time to coordinate the suck/swallow/breathe cycle.

Hand expression is a technique that usually, but not always, allows us to remove some milk from the breast.

If no colostrum is collected via hand expression, it does not indicate that milk is absent in the breast.

It may indicate that the hand expression technique used is suboptimal, but it could also be a function of timing during the First 100 Hours.

It is still important that it is attempted because the massage and compression of the breasts is a crucial part of the process of early lactation.

We can approach this situation by validating that hand expression has been done this time around.

If it needs to be done again, we start fresh and assume it will yield milk.

Remain positive and encouraging about hand expression.

Consider every session of hand expression to be an opportunity to practice the technique.

Expect and teach that in the earliest hours of lactation, hand expression sometimes yields milk and sometimes does not.

Breastfeeding should still be continued regardless of whether any amount of colostrum is collected from hand expression.

Outcomes of hand expression or pumping do not accurately reflect whether or not a baby is transferring milk from the breast, and important transfers of microbes between the mother's skin and the baby's mouth continue to occur every time a baby is at the breast.

Encouraging hand expression, like skin to skin, is a first-line, high-priority, repeatable step that ensures we are doing everything possible to support our client to exclusively breastfeed and feed their baby with their own milk.

It's not a one-time task.

We must not facilitate skin to skin only in the "golden hour" right after birth and say "All good, we've checked it off," never to return to it.

We must not try hand expression just once to "see if there's milk" or to give the baby extra drops just one time.

These are repeating interventions that can be done every single time that we suspect or know that there's an issue with breastfeeding.

Only after these techniques—key pieces of the breastfeeding process —have been taught, demonstrated, and utilized should lactation tools or devices and infant formula be considered.

If these techniques have not been incorporated into the dyad's feeding routine, and we jump straight to introducing tools and devices, then we are effectively becoming product specialists—and promoters—rather than process specialists.

Being a breastfeeding process specialist also means continuing to encourage and promote feeding at the breast even when breastfeeding problems arise or are suspected.

There is rarely a true reason to stop breastfeeding completely.

Hypoglycemia

If a baby has low blood sugar in the earliest hours after birth, skin to skin prevents them from spending calories or burning fat stores trying to keep warm.

Hand expression offers a way to get more milk into them even if they are unable to feed at the breast.

If glucose gel is used to assist the baby in improving their blood sugar levels, it can easily be administered with the baby skin to skin and does not need to be considered an interruption to breastfeeding.

Jaundice

If a breastfed baby develops jaundice, skin-to-skin contact and hand expression are the perfect companions for the increased support the baby now requires with latching and positioning.

If phototherapy is needed, the Academy of Breastfeeding Medicine's protocol for managing jaundice suggests that the evidence supports interrupting phototherapy for breastfeeding for up to 30 minutes at a time.

If a phototherapy blanket device is used, breastfeeding can continue with no interruption of phototherapy.

When using a phototherapy blanket, mothers should be assisted in finding a comfortable position for breastfeeding, as the blanket can be challenging to manage.

A side-lying position may be very helpful in this situation; the mother and baby lie on their sides facing one another while feeding with the baby remaining on the phototherapy blanket.

Weight Loss

If a breastfed baby loses more weight than expected (generally more than 10% of birth weight) in the First 100 Hours, the plan must be to improve breastfeeding and get more milk into the baby. Skin to skin and hand expression are the primary strategies to complement the increased level of breastfeeding support and assistance now indicated.

Ensuring that pasteurized donor human milk is available in the hospital to any baby who needs it protects the infant's gut microbiome when hand expression does not yield enough additional milk to meet the baby's requirements.

Most often, needing additional milk in addition to breastfeeding and the mother's own expressed milk is a temporary and self-limiting situation.

Engorgement

While engorgement is often thought of as a common landmark of early lactation, it's actually a problem that requires attention.

Let's look at the timeline of the onset of milk production using a story.

Lanelle is pregnant with her first baby.

Around halfway through her pregnancy, hormones signal Lanelle's breasts to begin making milk.

However, those signals also indicate that the baby has not yet been born, so other hormones act to suppress copious milk production.

After Lanelle gives birth to her baby, she also gives birth to the placenta that nourished her baby.

Once the placenta is no longer in her body, new hormonal signals tell her breasts to increase milk production.

The process takes somewhere between 48 and 72 hours.

During that time, the milk is thick and concentrated, and Lanelle is taught to call it colostrum.

At some point after her breasts begin making a lot more milk,

Lanelle will feel that her breasts are more full.

This is key: Lanelle will not feel the increased milk production at the moment it begins to happen.

It may take many more hours for her to notice any changes.

Thus, first milk production increases (and therefore more milk is available to the breastfeeding baby when they are at the breast), and then later the mother feels changes in her breasts.

If no one teaches Lanelle about this, she may not realize that her baby is already getting more milk before she feels the changes in her breasts.

It is possible, though, that Lanelle will notice that her baby is swallowing audibly and more frequently and has more wet diapers before she even notices the breast changes.

With a normal, responsive, baby-led breastfeeding routine, Lanelle can expect that her breasts may feel fuller between nursing sessions and softer after the baby has nursed.

Lanelle's milk production will ultimately synchronize with the amount her baby is taking out.

As breastfeeding continues, her milk production will continually align with the signals breastfeeding provides to her brain.

Next, let's contrast this with Amy.

Amy is also breastfeeding her new baby, who was born 48 hours ago.

Breastfeeding has been going well and everything is progressing as expected - until she begins to feel breast changes.

Within hours, despite her baby continuing to nurse whenever they ask, her breasts are becoming increasingly full.

They become uncomfortable and it soon becomes difficult to latch the baby because the fullness flattens her nipples out somewhat.

Because Amy can't get her baby to latch well and nurse, she has to begin pumping and feeding her baby with a bottle.

Amy struggles the first few times she pumps because milk isn't flowing easily.

She wonders if she should feed her baby some formula until she can work this out.

Fortunately, Amy gets help from a postpartum nurse who helps her with gentle breast massage, a special technique called reverse pressure softening, and hand expression before she brings the baby to the breast.

This helps the breast to be more pliable so that the baby can latch and remove milk.

The combination of reverse pressure softening and hand expression can also help her milk flow easily if she needs or decides to use a breast pump while engorged.

Amy's nurse explains that sometimes this type of engorgement occurs when there have been a lot of IV fluids given to a mother during labor, such as when she has epidural anesthesia or a cesarean birth.

Amy learns from her postpartum nurse to watch the baby's swallows and track his diapers to know he's getting enough when breastfeeding.

Once home, Amy gets support via telehealth from a lactation consultant, and it takes a few days of breastfeeding, pumping, bottle feeding, and extra work to get her baby to breastfeed again.

When she has resumed exclusive breastfeeding and her baby has normal diaper output and appropriate weight gain, Amy washes her pump parts and puts her pump away until she is closer to the time she will return to work outside her home.

Amy has gained confidence in her ability to breastfeed her baby despite some early challenges, and she did not change her lactation journey permanently when experiencing a problem in the earliest days.

Though Amy's lactation situation became complicated when engorgement made it difficult to latch her baby and threatened to interfere with her milk production, Amy got the right information and help at the right time.

Amy is fortunate to have had a postpartum nurse who understood how milk production unfolds in the First 100 Hours and what to do to resolve common problems.

If any other situation arises where Amy wants or needs to use her breast pump, she's already had education about how to use it.

It shouldn't be a matter of good fortune, though.

Everyone who cares for breastfeeding dyads should have a thorough understanding of how early milk production progresses and what to do if problems arise.

Special Solutions For Early Or Unwell Babies

This book has so far primarily covered lactation education and solutions for breastfeeding a healthy, term newborn.

Next, we will look at babies who are born early (including the Late Preterm population at 34-36 weeks gestation) and/or are unwell at birth and require medical care.

Using The First 100 Hours approach continues to have immense value in these situations.

However, instead of relying on responsive breastfeeding to synchronize milk production to the baby's needs, in these special cases we must create the rhythm of milk expression through a routine of hand expression and, later, if desired, pumping.

This routine and rhythm provide needed feedback to the brain that though no baby is suckling at the breast, milk is still needed and the body should continue to use energy to produce it.

Without regular milk removal and in the absence of a baby suckling at the breast, the body will conserve resources by decreasing and then stopping the production of milk.

Because babies, on average, communicate that they need to go to the breast at least 8 times every 24 hours, the milk removal routine should mimic that pattern.

The type of milk expression that is best for this time frame may vary.

Since hand expression is a necessary skill for breastfeeding, it makes sense to introduce it early and practice it often.

Mothers with access to breast pumps (manual or electric) also need to be shown hand expression and encouraged to incorporate it into their daily milk expression routine.

It can be done before, during, after, or instead of a pumping session.

Mothers who are using breast pumps should receive thorough teaching and assistance to ensure they are comfortable with everything they need to know in order to use them.

They should be assessed to ensure they use pumping equipment, especially flanges, that is specifically fitted to their body and their needs.

Mothers who rely on breast pumps to maintain their milk production for any reason should be encouraged to use breast pumps manufactured for frequent, everyday use.

Breast pumps made for occasional, light-duty use are not efficient enough to maintain milk production when relying on milk expression in the absence of breastfeeding.

If an early or unwell baby is also breastfeeding, mothers should be taught to continue their milk expression routine as a backup method in case the baby is not yet efficient at milk removal.

It is important to educate new mothers with early or unwell babies on how milk production is expected to progress throughout the First 100 Hours.

We must help them set appropriate expectations for how much colostrum to expect to hand express in the earliest hours, how soon they might feel breast changes, and when they can expect higher volumes of milk when they are expressing.

Without this information, mothers may become discouraged about the small amounts of milk they are expressing, which can lead to disruption of their milk removal routines and poor breastfeeding self-efficacy.

Continuity Of Care

With the First 100 Hours approach, mothers are thoroughly informed, equipped with the skills and techniques they need to manage the earliest hours of breastfeeding, and prepared to ask for help if they notice breastfeeding or milk production outcomes that are not what they were expecting.

Ideally, within the First 100 Hours, every healthcare provider who has contact with the new mother and baby should be alert to any indicators of potential future breastfeeding problems, especially if the mother has reached 96 hours postpartum and has not yet experienced breast changes or an increase in milk production.

They should also be alert for other indicators of potential breastfeeding problems, such as:

- risk factors for low milk production
- insufficient glandular tissue
- infant oral anomalies like restrictions in tongue movement, inability to form a seal at the breast when latching, cleft lip and/or palate
- maternal perception of or concerns about low milk production
- any other potential factors that could create breastfeeding challenges in the future

If risk factors for breastfeeding problems are observed, especially if no milk volume increase is noted by 96 hours, the mother should receive a prompt referral to speak with an IBCLC.

All mothers should receive referrals to local, virtual or in-person, peer breastfeeding support groups in their area.

Licensed Baby Cafés are an evidence-based model of breastfeeding support which combines professional IBCLC care and peer support in communities across the United States; published data from Baby Cafés demonstrate that mothers who attend meet or exceed their breastfeeding goals.

Many other peer-led breastfeeding support groups have been demonstrated to offer excellent support to mothers.

These groups should be facilitated by knowledgeable facilitators or leaders who are trained to refer mothers and babies with medical or complex lactation needs to the higher level of care they require.

It is very important to note that peer breastfeeding support groups often provide more than just breastfeeding care; they can be a valuable resource where other health needs in postpartum women and new babies can be observed and referrals are given for appropriate care.

Suppose a hospital or birth facility does not offer outpatient lactation care services.

In that case, they should provide new mothers with resources for peer breastfeeding support groups and local IBCLCs who provide skilled lactation care.

Uncovering problems with breastfeeding and milk production early is key to resolving them as fully as possible.

Every interaction in the First 100 Hours needs to end with clear information about how and when to get additional help with breastfeeding to prevent small concerns from becoming bigger problems.

CREATING A BREASTFEEDING MINDSET

The First 100 Hours approach is designed to restore **trust in breastfeeding.**

Trust in breastfeeding has been eroded to a point where both parents and healthcare workers are unlikely to believe that it will be possible or safe.

On a population level, it is important that exclusive breastfeeding is understood to be a critical health behavior that is not only possible but optimal for most new dyads.

It is also crucial that it is treated that way; i.e. that healthcare workers always begin from the assumption that the dyads in their care will be able to breastfeed and that they provide care and support accordingly.

Trusting in breastfeeding is the logical approach because breastfeeding is the physiological human norm.

It is physiologically plausible that breastfeeding, rather than feeding a substitute for human milk, would be optimal.

The existing scientific evidence base demonstrates that it is possible and optimal for most humans.

An individual's trust in breastfeeding and the likelihood of accomplishing breastfeeding depend heavily on their own belief in their ability to breastfeed.

This is called breastfeeding self-efficacy, and it can be measured using a validated scientific tool.

It is uniquely constructed within an individual's self-identity, and it can be influenced by all of the people in a mother's life, including the healthcare providers around her and how they provide care and support.

For example, if a mother believes that she will be able to breastfeed easily and feels confident that she can get help if she needs it, she will likely achieve both of those objectives.

However, if she is cared for by healthcare workers who raise doubts frequently, her breastfeeding self-efficacy can be eroded.

Likewise, if a mother expresses that she is planning to try breastfeeding but feels unsure about her ability to do so, her level of confidence and trust in breastfeeding can be raised by receiving care from healthcare workers who trust in breastfeeding, treat it as the norm, and provide the practical support she needs to accomplish it.

I was fortunate to work with a postpartum nurse named Susan in the hospital.

She had such a relaxed and encouraging demeanor about breastfeeding that she became an important influencer on both her patients and her colleagues.

Susan was the nurse all the other nurses called into a room if they were having trouble helping a patient to breastfeed.

I had the opportunity to observe how she did it many times, and I found her approach incredibly powerful.

It almost seemed like the more challenged or anxious a new mother was feeling about breastfeeding, the calmer Susan became.

The moment she heard a patient was struggling to wake up the baby for feeding or to latch, she would immediately set about making the mother comfortable.

Using the bed controls to adjust the bed positioning, as many pillows as she could find, and rolled-up towels as needed, she could easily help a mother feel settled in, warm, and cozy with her baby on her body in skin-to-skin contact or in the side-lying position for breastfeeding.

She calmly settled both mother and baby into a position where breastfeeding could happen.

She would talk them through everything in her soft voice and reassuring way, explaining why the baby needed to be unswaddled and have their hands loose and how they could gently compress their breast to express a little milk to entice the baby.

Because she primarily worked the night shift, she was used to taking care of her patients' lactation needs without the lactation staff nearby.

If she noticed something that concerned her or recognized that there might be some red flags for breastfeeding problems, she would take responsibility for giving the patient a simple, gentle explanation of what she observed and what it could potentially mean.

She would also ensure that the lactation staff was notified so that we could follow up.

If she referred someone to us, we knew she had already assisted and observed breastfeeding, taught hand expression, and talked to them about the signs that breastfeeding was going well.

This helped us to know what we might need to do next for that patient.

Susan was a skilled nurse with training and experience supporting lactation, and she excelled because she understood that her role in lactation care during the First 100 Hours of a baby's life was to facilitate breastfeeding.

She created experiences of breastfeeding for the dyad so that they could build on what they learned and move forward even if there were problems.

As a postpartum nurse, the most helpful skills you can hone are what Susan did: latching and hand expression.

You don't have to solve all the breastfeeding problems you encounter; you just need to keep these actions going and make referrals for additional lactation care as needed.

Make all of it easier for yourself by always returning to skin-to-skin contact as the first step of the interaction with your breastfeeding patient, and then move into helping to latch the baby if help is even needed.

If the baby doesn't latch, proceed to the next step: hand-express and spoon-feed the milk that is expressed.

Keep it that simple, and you'll consistently give your patients confidence in these actions as fundamental to breastfeeding.

How healthcare workers "feel" about breastfeeding—how much they trust that it is safe, effective, and important—has a tremendous effect on the individuals for whom they provide care because it impacts the type of lactation care and support they provide.

Keep in mind that *your* truth (your lactation experience) is not *the* truth about how breastfeeding works.

It is critical that healthcare workers maintain a professional and scientific approach to breastfeeding, rather than allowing their personal experiences and biases to impact how they care for their patients and clients.

By empowering all new parents with the information they need to navigate the First 100 Hours of a baby's life, we can build their confidence so that they can continue to parent their newborn effectively and trust that they can reach out for assistance whenever needed.

Mothers should understand that breastfeeding is about knowledge and action, not magic, luck, or technology.

We often see social media posts where mothers tell their peers that they were "lucky that breastfeeding worked out."

While this might sound like a nice way to support mothers who struggle with breastfeeding, ensuring that they don't feel stigmatized or like they "failed" at something, one problem with that statement is that it minimizes the accomplishment of breastfeeding for the person who breastfed, which is also important, and it minimizes the support that the "lucky" one received.

One way we can support all the new mothers and parents we encounter is to encourage them to set much smaller and more specific goals than "I will breastfeed for six months" or "I will exclusively pump for a year."

By beginning with "just" the First 100 Hours, where important neural mapping is happening during breastfeeding and critical hormonal signals are being provided to the mother's brain, we present a meaningful and achievable goal for many.

For the mother who plans to pump her milk exclusively, we are supporting her in having optimal milk production right from the start.

For those mothers who find themselves needing to feed formula for medical reasons, we are optimizing all of the practices and techniques that will support continued breastfeeding and/or human milk feeding, and we have minimized the sense of formula "becoming" an ongoing feeding plan.

Here are the mindset shifts that can happen if a mother can successfully breastfeed her newborn or feed only her own milk during the First 100 Hours:

- Maternal confidence in breastfeeding grows exponentially = increased breastfeeding self-efficacy
- Memories of breastfeeding are imprinted permanently
- The value of breastfeeding to this mother increases
- A kinesthetic experience of the feeling of breastfeeding imprints for mother and infant
- The healthcare team surrounding and supporting the dyad gains experience and confidence in the process of breastfeeding

Help build a mother's confidence by praising what she has already been doing and building up her plan to include additional helpful techniques and information.

Mothers are able to respond better and engage more fully with you as their supporter when they perceive that their own needs are being met, so always check in on aspects of their recovery other than breastfeeding, even if your primary job is to provide lactation care or education.

Always validate a mother's concerns because every problem they perceive is important to them.

Provide the information they need to navigate lactation now and offer anticipatory guidance for what may come next.

Present step-by-step information about breastfeeding clearly and concisely.

Tailor the information to their specific needs and situation.

Gatekeeping information, or holding it back because you think they can't handle it, is unethical.

Clients can decide how much information they can handle – this is never a provider's right to decide.

Sometimes, your perception that they are "overwhelmed" is accurate.

Still, it is a temporary situation – so consider shortening your plan for them and having them follow up with you sooner to create another opportunity to provide support and information.

The First 100 Hours after a baby is born are understandably a challenging period for new parents.

However, they are also a crucial time for ensuring that relevant information is conveyed clearly and simply.

With the right information, babies and their mothers can thrive during this time, setting them up for an easier transition during the first few weeks of the baby's life.

The Value of Formula and Lactation Tools

Promoting breastfeeding and human milk feeding is about

nutrition, nurturing, and preventing disease.

We need strategies that protect the health of both infants and their mothers.

These strategies should also preserve breastfeeding in the short term, protect the mother's capacity for milk production, and facilitate the dyad's sustainability of breastfeeding over time.

Avoiding non-medically necessary infant formula is one of these strategies.

Delaying the use of non-medically necessary lactation tools and products is another strategy.

In the section of this book on the History of the First 100 Hours approach, I talked about how when I started working in the field of lactation care, formula use in the hospital was very common, but breast pump use was not.

During my career, I have watched breast pump use in the hospital become nearly as ubiquitous as formula.

Pregnant women bring breast pumps and other lactation tools and products to the hospital frequently.

While there are situations in which formula, breast pumps, and other lactation tools are medically necessary or indicated, the fundamental lactation needs of the mother and baby in the immediate postpartum period have not changed.

The major change that has brought these items into nearly every hospital postpartum room is marketing.

Mothers are primed (through the marketing messages targeted to them) to believe they will need to buy or acquire many things and that the earlier they start using them, the better off they will be.

Marketing messages have taught our culture that breastfeeding is hard, that many mothers can't make enough milk, and that it takes a lot of products to make sure a baby is fed well.

How an expectant mother feels about her ability to breastfeed (her breastfeeding self-efficacy) is a very significant factor in predicting whether she will be successful at meeting her goals.

A person with low breastfeeding self-efficacy is very vulnerable to messaging that indicates breastfeeding will be hard, unsustainable, and not valuable.

These types of messages are propagated constantly by companies that seek to profit from this vulnerability.

This is exactly how early breastfeeding becomes overly complicated.

Our role in early lactation support is to protect and preserve it by simplifying the process.

We must be specialists in the process of breastfeeding.

We must understand how formula and lactation products can assist in breastfeeding and how they can interfere.

We must consistently encourage the simplest and gentlest solutions to early lactation problems.

Formula and lactation products, including breast pumps, are interventions that must only be used when a problem is not resolved with better, more frequent breastfeeding and hand expression of extra milk to feed the baby.

In the First 100 Hours, the earliest hours of breastfeeding and milk production, it is critical to facilitate as many opportunities as possible for mothers to experience breastfeeding and learn to hand express their milk.

At the same time, these simple, gentle solutions promote more rest and recovery for the dyad.

We know that when a baby is born, their mother experiences a cascade of hormones that are designed to help keep her in a state where she is going to be focused on her baby's needs and her own body's need for rest.

It is a nurturing state.

It is a state where it is easy to focus on her baby.

It is a state where she is resting so that she can recover from childbirth.

Her brain functions optimally to respond to her baby in this oxytocin-rich environment.

Simple, gentle solutions like more frequent and optimized breastfeeding, skin to skin, and hand expression support this oxytocin-rich environment of nurturing, protection, and rest.

Introducing machines, tools, plastic items, products, smartphone apps to count how much milk is being expressed, and substitutes for mothers' own milk creates barriers to nurturing and rest.

They require different thinking, analysis, and decision-making that can override the nurture-and-protect mode and interfere with learning and experiencing how breastfeeding feels.

Many types of lactation products, tools, and infant formula can be useful in certain situations beyond the earliest hours and days of breastfeeding and lactation.

But most of them are not needed in the First 100 Hours.

Timely, effective lactation support and accurate information are the most important lactation tools in the First 100 Hours.

If a product, tool, or formula is actually needed, ensure that appropriate education is provided regarding how to use it and for how long, and encourage the mother to continue breastfeeding frequently and responsively and to hand express her milk to feed to the baby.

Offering this type of specific and thorough information when a breastfeeding mother needs to use a lactation product or formula temporarily demonstrates the value of these tools.

If a mother is advised to use formula to feed her baby, but she does not receive complete information on what problem it is solving, how much to feed, how often to feed it, how to know when to stop, and how to sustain breastfeeding while formula is being used, she will find herself combination feeding or exclusively formula feeding without fully informed consent.

Likewise, if a mother is advised to use a breast pump, nipple shield, silicone milk collector, or any other tool without explicit information about how to do so, her lactation care is incomplete.

She will find herself on a different feeding plan than she intended, again without fully informed consent.

Lactation products and tools often seem to multiply in a mother's feeding journey.

Much like the concept of the cascade of birth interventions, where one intervention leads to many more, there can be a cascade of lactation products whose use leads to the need for more and more equipment to maintain lactation.

It's often said that parents deserve choice when it comes to infant feeding; usually, this is said in relation to our obligation to support parents who want to feed their babies formula.

However, facilitating a mother's fully informed choice about how to feed her baby also means that we need to ensure that people who want to breastfeed have the care and support they need to do so.

It also means we must work to proactively counter the messaging mothers face from commercial influences who want them to purchase and rely on their products.

Our role as healthcare providers is to support practices that encourage and facilitate breastfeeding and discourage things that can interfere with it.

When working with breastfeeding mothers in the First 100 Hours, we rarely need to introduce formula or a lactation tool into a breastfeeding situation.

We must remember that when we do find that formula or a lactation tool are indicated, we are clear about discussing it as a temporary tool that can be discontinued and clearly inform the mother that using it is her choice.

Likewise, if a baby requires formula for a medical issue that is considered a necessity, the baby's mother needs extra support to ensure that she can maintain her confidence in breastfeeding.

The value of infant formula and lactation tools comes from ensuring that mothers are educated on when and how to use them appropriately and that they are not influenced to use them simply because they are commonly used by many mothers.

In the earliest hours and days of lactation, positioning formula in particular as a tool to solve breastfeeding concerns is problematic. Because it is so commonly used, its use is generally viewed as insignificant and its potential health consequences are minimized.

Infant formula manufacturers do everything they can to make brand-new parents believe that formula will always be easily accessible to them.

When hospital staff ask patients whether they want to "breastfeed or bottle feed" or "breastfeed or formula feed," they are positioning these choices as equivalent and minimizing the chances of patients asking questions or bringing up concerns.

Asking patients what they have heard about breastfeeding can help start a conversation that can reveal what they feel, believe, and need to know about breastfeeding.

This line of inquiry may sound like it leaves no space for mothers who do not want to breastfeed.

Still, in 10 years of working in the hospital, I can assure you that the relatively small percentage of mothers who do not want to breastfeed at all are very comfortable saying so because their decision has already been made.

Statistically, most mothers do want to breastfeed at least once, and many use the word "try" when they are asked if they will be breastfeeding.

Asking what they know about breastfeeding helps to reveal their needs, questions, concerns, and doubts.

It gives us a place to enter the conversation with information that is relevant to them and meets the needs they have just identified.

When someone WANTS to breastfeed, then supporting them to do so is our responsibility.

Breastfeeding is simple.

Our feelings, belief systems, and structural barriers make it complicated.

Early lactation strategies often require process-based solutions or techniques rather than products or tools.

The first strategy for early latch difficulties is adjusting infant positioning and latch technique, not a nipple shield.

The first line strategy for a newborn who requires additional intake is colostrum expressed by hand, not formula.

The first line strategy for early nipple trauma is improving latch and using expressed milk to heal nipples, not nipple cream.

The first line strategy for expressing milk in the first hours and days of life is hand expression, not a breast pump.

If a baby requires additional intake in the First 100 Hours and is given formula without any attempts at hand expression, their mother will not have learned about or practiced hand expression should they experience engorgement or not have access to a breast pump later on.

If a mother is given a sample of nipple cream as a response to her report of nipple damage or pain, the root cause of the problem has not been addressed, and the damage and pain will likely continue. When a breastfeeding problem reveals itself, it's important to remember that there are multiple factors to be addressed: the baby's well-being in terms of milk intake, the mother's ability to continue or sustain breastfeeding, and the ability to sustain milk production over time.

Formula is often used as a tool to solve breastfeeding problems, but the only factor formula addresses is the question of milk intake, leaving the rest to chance.

Often overlooked is that formula has different physiological effects than human milk on the infant, and human milk is the optimal milk to be fed when a baby needs more than they are getting at the breast.

Safely prepared infant formula is indicated only when no human milk is available for these reasons:

- a mother has chosen not to breastfeed or feed her own milk
- breastfeeding and hand expression have already been implemented during this feeding session and no additional expressed milk is available
- pasteurized donor human milk is not available

In the First 100 Hours of a baby's life, infant formula use should be treated as a temporary intervention (except in the case of maternal choice as described above) which may be discontinued at any time.

If formula is recommended, ordered, or suggested BUT no specifics are given as to when to stop, mothers will reasonably assume they should keep giving it.

If formula is medically indicated, mothers must be informed about:

- the specific plan being recommended, including: how much

formula to give baby, when to give it, and for how long (i.e., how many times)

- how to continue breastfeeding and protect their milk production when formula is used: this is necessary whether formula is being given instead of breastfeeding OR in addition to breastfeeding
- recommendations to use ready-to-feed formula whenever possible and on how to safely prepare powdered infant formula; powdered infant formula is not sterile and requires hot water to be used to prepare it
- how they will find out or know when they can stop giving their baby formula

Encouraging mothers to turn to - or to rely on - products or tools instead of their own bodies and abilities during the earliest hours of lactation erodes their confidence in their ability to care for their own child.

It also leaves them critically under-prepared to manage their lactation problems later on.

While there may rarely be occasions when formula or lactation tools are needed in the First 100 Hours, those situations should be considered unusual.

Treat these unusual situations with the appropriate level of importance by providing comprehensive education and assistance to use the products properly and only as needed so that additional problems are not created through their use.

Breastfeeding, skin to skin, and hand expression are simple, sustainable practices which can be used continuously throughout the First 100 Hours and beyond, solving many of the most common challenges of early breastfeeding.

ACHIEVING THE VISION

How will we know when the First 100 Hours approach is working?

- When we hear mothers asking their babies' pediatricians, "What if I breastfeed more and hand express my milk instead of giving some formula?"
- When postpartum nurses everywhere share with us how, with their encouragement and help, a mother hand expressed for 3 feedings in a row and then was able to get her baby to latch and have a feeding at the breast
- When rates of non-medically necessary formula supplementation in the hospital decrease
- When attendance numbers at community-based, peer breastfeeding support groups blossom
- When mothers report that they are breastfeeding much longer than they thought they could or would
- When more babies get more human milk for longer
- When mothers and babies experience the long-term health protections that breastfeeding confers on them
- When mothers express gratitude that they were able to make milk for as long as they wanted and to end breastfeeding when they and their babies wanted to
- When mothers report that they never felt forced to feed their babies formula or human milk

- When all hospitals hire non-RN IBCLCs, certified lactation educators and lactation counselors to reach full staffing and ensure every new mother gets the lactation care she needs
- When IBCLCs are recognized worldwide for their expertise in breastfeeding and human lactation and for their full support of the WHO Code
- When mothers talk about how they were supported and educated about breastfeeding by everyone around them so that they could make their own choices without commercial influences
- When everyone in the lactation field is resilient, adaptable, and healthy as they grow and mentor the next generation of lactation care providers

LEADERSHIP AND ACTION IN THE LACTATION FIELD

T here's a common saying in the lactation field: when you're frustrated with the system, just focus on helping the one mother and baby in front of you at the moment.

It does help you center yourself, and it does matter.

If you're reading this book though, it's pretty likely that you're feeling motivated to work on improving the environment of lactation care on another level as well.

This is the section for you.

Let's revisit the vision and mission of The First 100 Hours approach.

Vision: New mothers and babies thrive in the First 100 Hours after birth

Mission: By providing accurate, evidence-based information about human lactation and infant development, new mothers are guided with appropriate expectations about newborns and supported with the skills and techniques they need to optimize breastfeeding and health within the First 100 Hours and beyond

Break that down even further: **the essential mission is that every mother has the support she needs to breastfeed.**

Remember, we're not making a choice on her behalf, and we're not persuading her: we're giving her the **information, techniques,**

and practical knowledge she needs in order to breastfeed if she so chooses.

If even one of those things is not provided to her, she really doesn't have the choice to breastfeed.

She might be able to figure it out, but that's going to be much harder on her than if she's supported.

Now that we all understand the mission, it's time to execute it.

Believing in the mission doesn't make it happen.

You have to drive the change and move the mission forward.

The First 100 Hours IS the strategy to make this mission a reality.

We need more than strategy: we need steps and actions to implement the strategy and fulfill the mission.

Here are our 4 Steps: Plan, Simplify, Connect, and Empower.

These steps will constitute a cycle of action for you.

You will always be moving within the steps of this cycle - that's how you will mark your progress.

Plan

When you plan, you will be considering and determining what elements need to change in your practice setting in order to fulfill the overall mission.

One question that may help you plan is to ask yourself, "What if it were easy to support the mothers giving birth in our hospital to just breastfeed and feed their own milk?"

Think about what would need to be in place, who would need to be involved, and what that could feel like for your patients.

It might involve creating a new policy or implementing one that has not been implemented fully.

It could revolve around providing a specific piece of information to staff so that they can relay it to patients.

Some elements of the plan will feel very small and easy to achieve, while others will feel monumental at first.

For example, if you know that your hospital needs to create and institute a pasteurized donor human milk policy, that is a large element that will take time.

You will be breaking that element into even smaller stages and using the rest of the steps to make progress.

Be sure to build a timeline for the change, and include information about how and when the outcomes will be reviewed and reported.

Connect

In this step, you'll be focused on people.

You will want to identify everyone who needs to be involved in order to implement change.

You must also identify the people who are most influential among the entire group you'll need to influence.

For example, if you will be educating a group of nurses on a new procedure that better supports breastfeeding progress, like delaying baths for babies, you will want to first talk to the nurses who seem to find change easy.

These "early adopters" will be accustomed to taking the initiative when something new is proposed; they may have lots of questions that help to clarify what needs to be done, and that's a good thing. If they are influential, then when others are wondering what the change is all about, they will be there to answer questions and make the change more palatable for their peers.

You are building relationships with the people who can make your changes possible.

Communicate clearly and consistently about the plan and its timeline so that there are no surprises along the way.

This type of relationship-building is critical for keeping your plans moving forward.

Your energy around the project will impact its potential for success, and you will represent the change to many of the stakeholders.

Take care of the people who matter in your project by being available, consistent, and transparent.

Make sure they have everything they need to be successful.

To trust in your project, they'll need to trust you.

Simplify

Every chance you get, in every way you can find, simplify things for others.

When you are educating mothers, keep things simple.

When you are executing a multiple-step project with your colleagues, simplify each step as much as possible so that it feels achievable.

When you are working on a long-term issue, continually simplify the parts that are becoming confusing or getting misconstrued.

In any situation where information needs to be communicated among many people, there is a high potential for it to lose its focus or its details.

If you are driving an element of change, be sure that you are personally communicating the same message to everyone.

For example, if your hospital has committed to hanging posters in every postpartum room to help educate parents about how to understand their baby's cues, think about why that project will make a difference.

Then drill down even further.

Sure, a poster in every room means that the hospital is providing more education to patients.

But it also means that it is providing the same education to every patient.

In simple terms, "we're using the posters so we can make sure all of our patients get the same information about breastfeeding."

When someone asks why there has to be one more thing hanging on the cluttered walls, or why they now are expected to point out the posters to all their patients, or why there's so much breastfeeding "stuff" going on lately on the unit - the answer is 'We're using the posters so we can make sure all of our patients get the same information about breastfeeding."

If they continue to ask why, that's when you can break it down and share with them how inconsistent messaging around breastfeeding is one of the top reported frustrations of mothers who began breastfeeding in the hospital.

You can expand on why the hospital is taking on more breastfeeding initiatives: remind them of the essential mission and tie it into the hospital's overall mission of caring for people's health.

Then return to the original, simple explanation: "We're using the posters so we can make sure all of our patients get the same information about breastfeeding."

This helps you to continually center the mission while also holding space for people to have questions, objections, or concerns about the element of change you are implementing.

Empower

We don't have to do it all ourselves, and we shouldn't.

When we have planned the elements of change, built relationships with stakeholders, and simplified our messaging consistently, we have the ability to empower others to join us in the implementation.

Everybody gets all the information.

People resist new initiatives when they worry that there's an agenda they can't see, so be sure that everybody gets all the information.

Ensure that everyone knows the essential mission and the elements of the plan.

Give them the opportunity to bring up potential problems with the plan so that they are invested in making it work.

Be generous with finding and creating resources - written, online, etc. - that will help them feel very confident in what they are doing going forward.

Build trust with them so that they know you are there to support them.

Then let them do the work.

Avoid the temptation to micromanage the change.

Honor your commitment to the plan, review outcomes as directed by the plan, and be available to support new needs that pop up from time to time.

In the meantime, keep building and nurturing the relationships you need and keep the mission at the center of your focus.

Shifting the culture of lactation support is possible, no matter where you currently practice.

The shift may feel incremental at times, but it is achievable.

Here are some ways that I believe we in the lactation field can be leaders in improving overall health by ensuring that mothers have the support they need to breastfeed:

- We can look around us to identify others who have expressed an interest or an aptitude for providing exceptional lactation care. Show them how to start the process of learning about lactation and becoming certified to provide lactation support or skilled lactation care. (Remember Susan, the wonderful postpartum nurse I described earlier? She's now also an IBCLC.)
- We can open our hearts and our minds to the opportunity to mentor others who are entering the field of lactation care. Mentoring comes in many forms, and it has a significant impact on the future of those who are being mentored. We can mentor the next generation of lactation care providers and guide them to thriving in their purpose - in fact, we are the only people who can.
- We can advocate for hospitals to hire non-RN IBCLCs. This would increase the hiring pool of highly-qualified and skilled IBCLCs and allow hospitals to increase and improve the level of lactation care their patients receive.
- We can advocate for hospitals to hire non-RN lactation educators and counselors onto their lactation staff so that those individuals can increase the hospital's capacity for providing fundamental lactation support while providing the opportunity for those individuals to gain the clinical experience they need to apply for the IBCLC exam.
- We can advocate for IBCLCs to be utilized to the fullest extent of their credential by ensuring that their job

descriptions reflect their role as consultants. In this way, employers can be good stewards of their resources, providing their patients with the appropriate level of lactation care needed for their situation and holding accountable other staff to support patients' fundamental lactation support needs.

- We can continually build our community networks to ensure that continuity of quality lactation care is a reality for the people we serve. Learn about who else offers lactation support in your community - IBCLC care, lactation educators and counselors, and peer breastfeeding supporters and groups - and build relationships with these folks so you can refer your patients to them.

- We can build bridges of communication and trust that will strengthen the public's trust in our work and advocacy. The dedication, passion, and commitment of lactation care providers are powerful forces that can shift policy toward improved health for families through breastfeeding and human milk.

Above all, I invite you to seize the initiative of improving the First 100 Hours and ensuring mothers have the support they need to breastfeed.

Be ready to adapt to the changing needs that our world presents.

Hold yourself to the highest professional standards so that it's easy to see that you are making progress.

Continue to learn and marvel at the wonders of breastfeeding and human lactation every day.

I'll be right there alongside you.

PUTTING THE FIRST 100 HOURS APPROACH INTO PRACTICE

My website allows you to download PDF versions of all of these forms, as well as additional supportive resources. www.ChristineStaricka.com

Appendix A is a crib card that can be provided during prenatal education to educate mothers and their support persons about the First 100 Hours.

When their baby is born, they can use the card in their crib (perhaps in addition to the standard crib cards provided by most hospitals to ensure that their infant feeding preferences are respected.

(100)   **My First 100 Hours**   (100)

My Name Is --

I was born on _____ at_____ a.m./p.m.

I will be 100 Hours Old on _____ at _____ a.m. / p.m.

Please keep me **skin to skin** most of the time

I will be breastfeeding **as often as I want** for as long as I want

If I need extra milk, please help with **hand expression** and feed me with a **spoon or cup**, especially during my **First 100 Hours** of life!

(100)   Thank you for taking care of me and my family!   (100)

Copyright©2024 Christine Staricka, IBCLC || www.first100hours.com

Appendix B is a timeline and to-do list for the new mother in the First 100 Hours. It can be provided prenatally or once the baby is born and used as a reminder of the activities to be prioritized during this period.

To-Do List for First 100 Hours of Life

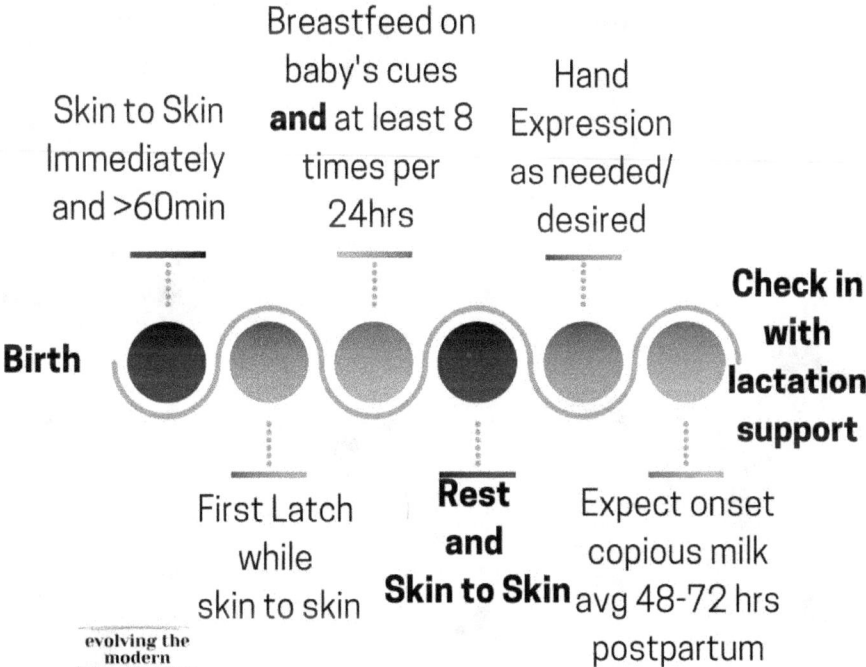

Breastfeed on baby's cues **and** at least 8 times per 24hrs

Skin to Skin Immediately and >60min

Hand Expression as needed/ desired

Birth

Check in with lactation support

First Latch while skin to skin

Rest and Skin to Skin

Expect onset copious milk avg 48-72 hrs postpartum

evolving the modern breastfeeding experience:

birth
0-24
25-48
49-72
73-96
100 hours

holistic lactation care in the first 100 hours

CHRISTINE IBCLC
Staubke

@IBCLCinCA

Appendix C is a guide for those providing bedside care of breastfeeding couplets. It contains useful tips and guidance for simplifying lactation support and tips for counseling new mothers on breastfeeding.

Bedside Lactation
strategies

Simplifying Early Lactation Care

When providing lactation care in the earliest days after birth, it can easily seem complicated and challenging. This lens focuses on what is most important during the First 100 Hours of a baby's life, illuminating the critical behaviors and techniques that can set new mothers up for success in meeting their own infant feeding goals with a clear understanding of the timeline of milk production.

These bedside strategies, combined with thorough observation and assessment of breastfeeding outcomes, will help you as a provider to feel organized and confident that you have met the needs of the new family during this brief and brilliant time of their lives.

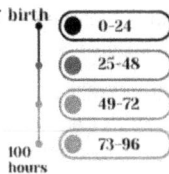

evolving the modern breastfeeding experience:

holistic lactation care in the first 100 hours

birth
0-24
25-48
49-72
73-96
100 hours

CHRISTINE IBCLC
Staricka

Important

These strategies address the needs of the healthy, term newborn. When working with babies outside of that population, additional interventions and strategies may be needed.
This should not be construed as medical advice.

Bedside Lactation
strategies

These 5 tips will help you strategize how to assist and educate parents at the bedside using best practices for optimizing early lactation!

☐ **Facilitate skin to skin** every chance you get. Remind mothers that skin to skin is where babies are designed to be and that it will help them breastfeed more easily.

☐ **Talk about responsive feeding** - letting baby tell them how often and how long to feed - as a way to make sure baby gets enough milk. Babies go to breast for many reasons, and food/nutrition is only one of them.

☐ **Talk about the hormones** that are being released inside mother's body whenever you see her baby skin to skin or feeding at the breast. Let her know that these hormones are signaling her brain to make more milk for her baby.

☐ **Hand express milk** and spoon feed if baby needs extra milk or isn't latching. This is a good opportunity to explain how colostrum is thicker and more concentrated and how she will soon see milk that is more watery and flows out easier.

☐ **Focus on assisting with breastfeeding early and often.** Babies will not breastfeed more often or longer than they need to.

Bonus

- Because babies are abdominal feeders, use the laid-back or side-lying positions as a default or if there are latch issues. **Ensure the baby is really well-supported** so that they feel physically stable.
- **Babies won't latch if their head is pushed forward**, and they won't stay latched if their chin is tucked toward their chest. Remind parents to keep their fingers off the back of the baby's head.
- If a mother seems frustrated about what you know to be healthy newborn behaviors, **praise the baby directly** for doing what they need to do while sympathizing with the mother and offering simple, breastfeeding-supportive ways to manage their challenges

Appendix D is a Feeding Log for the First 100 Hours. Checkboxes make it easy for new mothers and their support persons to track their baby's feedings and diapers in the First 100 Hours. It offers clear guidance and reminders on the minimum number of feedings, wet diapers, and stool diapers that they should expect in each of the 24-hour periods of the First 100 Hours, as well as what to do if they suspect a breastfeeding problem.

Feeding
Log

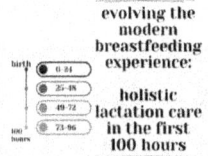

evolving the
modern
breastfeeding
experience:

holistic
lactation care
in the first
100 hours

CHRISTINE IBCLC

Keep It Simple

Hold your baby **skin to skin** most of the time in their early days. This allows them to access the breast every time they need it, reduces their stress hormones, and keeps them warm so they don't waste energy.

While skin to skin, your baby may signal that they need to be on the breast by bringing their fists to their mouth, opening their mouth, turning their head to the side if something touches their cheek, and making tongue movements. **These cues always mean that they need to be on the breast.** Babies who are held skin to skin more will be able to show these signals easily and more often.

More early breastfeeding results in better milk production and more intake by baby.

You can bring your baby to the breast every time they ask, and you should let them nurse for as long as they want. Sometimes they will fall asleep at the breast, and other times they will not. Both are normal. It's important to allow baby to be in charge of how often and how long they nurse because this is how they get enough milk.

If you stop them or limit their access to the breast, they may not get all the milk they need.

If your baby isn't latching or needs extra milk for any reason, **hand express** milk into a spoon and feed it to baby. If there is a feeding session where baby does not latch or only latches for a few minutes, the **spoon feeding** will provide them the milk they need for that session.

Next Steps:

Reach out for additional help and support with breastfeeding along the way

This is not medical advice.

1st 24 Hours of Life -- 8 or More Feedings

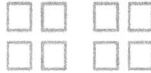

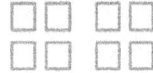

1 or More Wet Diapers
1 or More Dirty Diapers(Black)

2nd 24 Hours of Life -- 8 or More Feedings

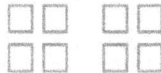

2 or More Wet Diapers
2 or More Dirty Diapers(Black)

Feeding
Log

evolving the
modern
breastfeeding
experience:

birth

0-24

25-48

49-72

73-96

100 hours

holistic
lactation care
in the first
100 hours

CHRISTINE IBCLC
Staricka

Keep It Simple

Your body has been **making milk** since about halfway through pregnancy, and now that your baby has been born, your body knows it should start making more.

Sometime during the First 100 Hours you will notice that your milk production has increased, your milk is thinner and whiter, and your breasts may start to feel different.

If your baby is latching, having at least 8 feedings every 24 hours, and having at least the minimum number of diapers shown here on the log, you will most likely not need to use a breast pump.

If your baby is not with you and you need to use a breast pump, **gently massage your breasts before pumping** and hand express when your pumping session is over.

If you are pumping **after** your baby has been feeding at the breast, pump only until your breasts feel softer and milk flow slows down.

If you are pumping **instead of** feeding your baby at the breast, pump for around 20 minutes and hand express before you're done.

Ask for help with latching, hand expression, breasts that become overly full, or any other problems or challenges that may arise. **Getting help right away is extremely important.**

Next Steps:

Reach out for additional help and support with breastfeeding along the way

This is not medical advice.

3rd 24 Hours of Life -- 8 or More Feedings

3 or More Wet Diapers
3 or More Dirty Diapers(Black-Green)

4th 24 Hours of Life -- 8 or More Feedings

4 or More Wet Diapers
3 or More Dirty Diapers(Green-Yellow)

Appendix E, The 3 Best Questions in the First 100 Hours, is a helpful guide for anyone providing lactation support to ask carefully scripted questions specifically designed to garner the most relevant information from your breastfeeding patient/client so that you can assess and evaluate their breastfeeding progress and determine their needs.

3 Best Questions About Breastfeeding in the First 100 Hours

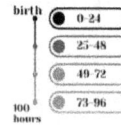

birth ● 0-24
25-48
49-72
100 hours ● 73-96

evolving the modern breastfeeding experience:

holistic lactation care in the first 100 hours

CHRISTINE IBCLC
Staricka

Don't Ask This	Ask This Instead

1

How often does your baby feed?

How do you know when your baby needs to go to the breast?

Encourage client to share with you **the basic timeline of their baby's feedings;** this shows you what they know about how babies cue to go to the breast

2

How long does your baby normally feed?

How can you tell when your baby is finished at the breast/chest?

Encourage client to give you information about **their own observations** of how their baby behaves at the breast and talk about **what they have heard** about how long babies "should" feed

3

How is breastfeeding going for you?

How do your breasts feel when your baby comes off?

Asking **how it feels** takes the focus off **volume of milk** and places it on **the act of breastfeeding**, allowing a better insight into technique, additional needs, and general progress

3 Best Questions for Exclusive Pumpers in the First 100 Hours

evolving the modern breastfeeding experience:

holistic lactation care in the first 100 hours

CHRISTINE IBCLC
Staricka

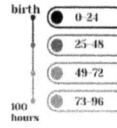

birth
0-24
25-48
49-72
73-96
100 hours

Don't Ask This	Ask This Instead
1 How often are you pumping?	**When was the last time you expressed milk?** Encourage client to think about the **actual times** of the last few milk expressions, giving you a sense **how they are timing their sessions** and whether they are hand expressing or pumping
2 How long do you usually pump or hand express?	**How long does it take to get all the milk out when you pump or hand express?** Emphasize the **act of milk expression** rather than milk volume; allow for more information to flow about **milk expression technique & milk production**
3 How much milk are you getting out?	**How are your breasts feeling after you are done expressing milk?** Asking **how it feels** takes the focus off **volume of milk** and places it on **the act of milk expression**, allowing a better insight into technique, additional needs, and general progress

Appendix F is The First 100 Hours Evaluation worksheet,

which provides an organized way to collect and document the information you collect by asking the 3 Best Questions (Appendix E). By looking at the information collected on the worksheet about what has already happened for a mother and baby and their breastfeeding history, it is simple to see what is left to do for them in terms of educating and assisting them with learning the techniques they need to sustain breastfeeding.

When used in combination with the First 100 Hours Feeding Log (Appendix D), a very clear picture should emerge of how breastfeeding has already been supported, the outcomes that have been achieved, and what can be done to support the mother and baby at this time.

First 100 Hours
evaluation worksheet

Mother's Name ....

1st Baby? **Yes/No**

Baby's Name ....

Born at .... Weeks ....

Birth Weight ....

Small for gestational age? **Yes/No**

Baby's Birth Date & Time ....

Date & Time Baby Will Reach 100 Hours of Age

....

Date & Time of
Consult ....

Feeding Observed **Yes/No**

Birth & Breastfeeding History

- Induction of labor
- Cesarean birth
- Preterm or Late Preterm birth
- Separation of dyad
- Has latched well at least once
- Used pacifier
- Fed formula
- Initiated pumping
- Used nipple shield
- Used bottle

Additional Factors of Note

- Nipple or breast pain
- Latching problems since birth
- Inadequate diaper output
- Jaundice
- Hypoglycemia
- Weight loss beyond normal limits for age
- Used supplementer at breast
- Engorgement
- Phototherapy
- Infant remains hospitalized >100 hours

First 100 Hours
evaluation worksheet

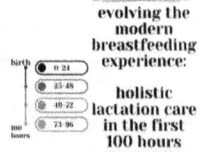

evolving the modern breastfeeding experience:

birth
0-24
25-48
49-72
100 hours
73-96

holistic lactation care in the first 100 hours

CHRISTINE IBCLC
Staricka

Education Provided

- ☐ Latch & Positioning
- ☐ Hunger/Satiation Cues
- ☐ Normal Newborn Feeding Patterns
- ☐ Timeline of Milk Production
- ☐ Expected Diaper Output
- ☐ Hand Expression
- ☐ Spoon/Cup Feeding
- ☐ Care of Sore Nipples/Nipple Trauma
- ☐ Breast Massage & Reverse Pressure Softening
- ☐ Oral anatomy issues

Other Notes

Appendix G is Lactation Management Planning, an outline you can use when working with a breastfeeding mother to create a plan for sustaining or improving her breastfeeding situation. The boxes and spaces can help direct you in ensuring that each piece of the puzzle is addressed and that the mother's goals are given attention and used as the basis of the plan.

Lactation Management
planning

evolving the modern breastfeeding experience:

holistic lactation care in the first 100 hours

CHRISTINE

Mother's Feeding Goals	Baby - Currently	Mother - Currently	Milk Production - Currently
Additional Planning	Baby - New Plan	Mother - New Plan	Milk Production - New Plan

ABOUT THE AUTHOR

Christine Staricka

Christine Staricka is a Registered, International Board-Certified Lactation Consultant and trained childbirth educator.

She developed the concept of The First 100 Hours approach to early lactation care when she worked as a hospital-based IBCLC for 10 years.

Christine has been working in the lactation field since 2001, providing clinical lactation care and support both in the hospital and the community. She has been the facilitator of Baby Café Bakersfield since 2014, which was the first licensed Baby Café to open in California.

Christine is the host of the Evolve Lactation Podcast and writes a blog on Substack called Evolve Lactation.

Christine is a Fellow of the International Lactation Consultant Association. She holds a Bachelor's Degree from the University of Phoenix.

She has been married for over 30 years, lives in California with her husband and dogs, and is the proud mother of 3 amazing daughters.

You can find more information about Christine and access free resources and downloads at www.ChristineStaricka.com.

ACKNOWLEDGEMENTS

To my parents, Frank and Lucille, whose love of reading and learning made me who I am today. Your instinctive knowledge of what it takes to raise and support a child to thrive, along with your strong belief in the importance of family and home, are just some of the many gifts you have given to me.

To my husband Jon, it is because of your encouragement that I wrote this book. Your ability to see things clearly in black and white improves my vision. Your support and love keep me grounded and honest. I thank you for all of it, and I thank you for helping me make our little family a true force in the world and for always, always, showing us the humor in everything.

To my IBCLC mentor, Lynn, who first guided me through early motherhood as the facilitator of our breastfeeding support group and then invited me into the lactation field. You changed my life in profound ways and gave me the gift of seeing breastfeeding as a fascinating, miraculous, and priceless part of life. You inspired my academic pursuits, my sense of advocacy, and my passion. Thank you for everything.

To Mary, my friend and colleague, you are a role model for every nurse and IBCLC. Your wisdom and patience have made me a better IBCLC. Your friendship has made my life richer. I will always treasure our work together, caring for mothers and babies and guiding them into their new lives together.

To Adrienne, my friend and colleague, you made me a teacher, and for that, I am so grateful. You told me I could teach and you taught me how to do it better. Your gifts as an IBCLC are incredibly unique, and your ability to make connections between people and information across disciplines is a valuable gift to the world. Thank you for always keeping the humanity at the center of what we learn and do together.

To Leslie, my friend and colleague, you have believed in me since the moment I met you. Your sense of wonder when you first really learned about lactation was awesome to watch, and it has never dulled in the years since. You have turned that wonder into a beautiful career and made a difference for so many new parents and babies. Thank you for always indulging me and encouraging me in going deeper and learning more.

To Cathy, my friend and colleague, you fully embraced every opportunity we have had to work together by making the collaboration itself the most important part. Your insights about people, mental health, and breastfeeding continue to inspire me to constantly seek ways to demonstrate care and sensitivity. You showed me how important boundaries are, and you taught me to take care of myself so that I could take care of others. Thank you for taking care of me.

To Tennie, my friend and colleague, you have modeled grace and love in your lactation and childbirth work. Your patience and care in supporting nurses is particularly inspiring; because of you, many nurses have carried the knowledge of lactation that you kindly, consistently, and lovingly imparted on them over years of nighttime shifts. Your deep empathy for everyone you meet leaves us all feeling cared for in the world. Thank you for showing us all that it's always ok to take extra time to care for mothers and babies.

To Lucia, my friend and colleague, your passion for mothers and babies is alive in every interaction you have, leaving us all invigorated and ready to get back out there and do our work. Your enthusiasm is contagious and productive. Thank you for keeping me close, showing me new opportunities, and being an absolute light in the world.

To my friend Danielle, your unwavering support has kept me moving forward in writing this book. You always say what I need to hear, and you can always get me to laugh even when I want to cry. Thank you for being there for me.

To my friends and colleagues Fleur and Fiona, your insights and perspectives keep me thinking, wondering, and creating. I am so grateful for your generosity in sharing your ideas and thoughts. From both of you I have learned what true, effective collaboration can look like. Thank you for challenging me to be true to myself in this work.

A special thank you to all of the IBCLCs I have mentored because it is from you that I learn the most. Your insights, observations, questions, and knowledge have undeniably changed my perspective and made me go back to learn more.

And thank you to everyone I have worked with, volunteered with, those who have hired me to speak for their staff, those who have sat and listened to me teach, and those of you reading this book. I hope you are inspired to act and to lead!